Thomas Wesley Mills

The Nature and Development of Animal Intelligence

Thomas Wesley Mills

The Nature and Development of Animal Intelligence

ISBN/EAN: 9783337026974

Printed in Europe, USA, Canada, Australia, Japan

Cover: Foto ©berggeist007 / pixelio.de

More available books at **www.hansebooks.com**

The Nature & Development

OF

Animal Intelligence

By

WESLEY MILLS, M.A., M.D., D.V.S., F.R.S.C.

PROFESSOR OF PHYSIOLOGY IN M'GILL UNIVERSITY, MONTREAL, CANADA,
AUTHOR OF "ANIMAL PHYSIOLOGY," "COMPARATIVE PHYSIOLOGY,"
"THE DOG IN HEALTH AND IN DISEASE," ETC.

LONDON

T. FISHER UNWIN

PATERNOSTER SQUARE

1898

PREFACE

From various quarters the suggestion has come to me to prepare a work on Comparative Psychology, as it was known that this subject has engaged my attention in no small measure for many years. It would be easy enough to collect an additional number of anecdotes of animals, and pen some reflections on them. It might be possible to gather together some accounts of the doings of animals of undoubted accuracy and examine these critically, but all this has been done, and we must now enter on another stage—that of exact, systematic observation and experiment. There are, however, many methods by which so broad a science as Comparative Psychology can be advanced, and I hope no word I may write may suggest any of those narrow views for which even scientific men are sometimes to be held responsible. There are many points of view, and it will be well to gather observations and opinions from every proper source available.

My own views as to the nature and scope of Comparative Psychology will best be gathered from the following pages, so that I will now quote to the reader a few lines from the writings of others.

Prof. C. S. Minot, in a review of Prof. C. Lloyd Morgan's "Habit and Instinct," writes thus :—

"As a naturalist, it has seemed to me that the naturalist's method has an immense future in Psychology. The method includes two main factors: the observation of details, and the comparison of homologous phenomena in different forms of life; and the method starts from the standpoint of evolution. There need be no restriction, of course, upon the three aspects of Psychology which have heretofore prevailed —the metaphysical, introspective, and experimental, but there should come soon, and with revolutionary power, not merely enlarged interest in, and sympathy with, Comparative Evolutional Psychology, but more than that—eagerness to enter this field of enquiry and to share in harvesting it" (*Psych. Rev.*, vol. iv. No. 3, p. 313).

Those who do me the honour of reading the pages of this book will learn for themselves how completely I share Professor Minot's views, and that my convictions have been followed by corresponding action.

But one may well ask: Who is able for so great a task? I know of no higher ideal of the requirements for the worker in Comparative Psychology than that set forth by Prof. Groos in his "Die Spiele der Thiere." He well says: "The author of a psychology of animal play should have, in reality, not alone two, but many souls within his breast." He would have him combine, with all the varied ideas and experiences of a man who has traversed the round globe, the special knowledge of the director of a zoological garden, and also that of him who has penetrated the life-secrets of the forest, and who can, moreover, take the point of view of the student of æsthetics. If these are the qualifications for the special investigation of animal play, they are not less called for in the other realms of Comparative Psychology.

While the present writer can lay claim to no such roundness of culture, he thinks he can confidently assert that nothing will be found in the following pages that has not some basis in his own observations or experience.

The Author has on more than one occasion expressed his belief that mere closet psychology is of little value in advancing the subject as applied to animals. Accordingly, it will be found that by far the greater part of this book is made up of the facts of observation.

In determining the form the work should take, I had to consider whether to re-cast all the material I had been accumulating for the last fifteen years, or republish what had already appeared in an almost unaltered form. It seemed to me that in the end the advancement of the subject would be best served by the latter course. While there may be some repetition in the papers that constitute the first part of the work, this will serve to emphasise the views that have been impressed more and more on one who has for ten years been in daily intimate association with animals, and a close and unprejudiced (as far as may be) observer of their life-ways.

Unless I mistake, there is now an interest in the study of animals altogether unprecedented, and I hope to see appear, within the next few years, accounts of researches which, in many respects, will be in advance of anything yet produced. It is largely with the view of stimulating such researches that I have concluded to publish the principal results of my own observations and thinkings up to the present time, in a form readily accessible to all who may be interested in such studies. As I believe the facts to be of by far the most value in our present stage of progress, they will be found to preponderate over theory. Laws, of course, we should try to establish, but I believe that we must go on as patient observers and experimenters for many years yet to come, before large generalisations can be safely formulated.

There will be some—possibly in high places—who may hold such work in light esteem, but they will find that, sooner or later, their lofty seats must be vacated, and that they must come down and delve. Filling up books and periodicals is one thing, and reaching truth another.

The work is divisible into four parts. Part I. consists of addresses, in which my own views of the subject are set forth. Part II. of studies, largely practical, of two interesting states—feigning and hibernation. These are on the borderland between natural history and psychology, but must have special interest from whatever point of view regarded. Part III. is, in my own opinion, much the most important part of the book. It is a storehouse of reliable facts, from which each reader may draw his own conclusions. Part IV. consists of discussions on a subject of considerable importance, as will be evident. This division of the work into parts rather than chapters permits of the reading of any one division of the work by those who may not desire to peruse the whole.

The prosecution of these studies has necessitated the breeding and rearing of a large number of purebred animals during the last ten years, and while it has involved considerable expense for one who has received no special favours from Dame Fortune, the work has been a source of pleasure, and, it is hoped, some psychological and biological wisdom also to myself; and if it leads to a truer and more complete study of the psychic nature of animals than heretofore, the Author will feel that his time, labour, and money have not been expended in vain.

I count myself fortunate in being able, by the kind permission of Mr T. Mann Jones, of Northam, England, to embody in this book, in the form of brief notes, some of the observations and reflections of so experienced and accurate a student of animal intelligence as he has proved himself to be. Mr Jones took the pains to write me, at considerable length, after reading my notes on psychic development, for which I am very grateful.

"Animal Intelligence and how to Study It," appeared, in the first instance, in *The Sherbrooke Examiner;* "Comparative Psychology," etc., in Appletons' *Popular Science Monthly;* "Psychology and Comparative Psychology," in *Science;* "Squirrels: their Habits and Intelligence," etc., and all the papers following, as far as page 276, in the *Transactions, Royal Society of Canada;* the "Discussion on Instinct," etc., in *Science.*

I desire to offer my thanks to the Editors and Publishers of the before-mentioned periodicals for their courtesy in readily assenting to re-publication of the papers referred to, as also to those contributors who have kindly allowed me to embody their views in this book.

As the discussions in Part IV. are printed just as they originally appeared, perhaps it may be well to state that Professor Morgan's views on the subjects under consideration may be found fully developed in his "Habit and Instinct," and Professor J. Mark Baldwin's in papers that have appeared in the *American Naturalist, Science,* and the *Psychological Review*, though such contributions cover a wider field than that traversed in the discussions to which reference is now made.

THE AUTHOR.

CONTENTS

PART I.

	PAGE
Animal Intelligence, and how to Study It	1
Comparative Psychology: Its Objects and Problems	17
Comparative Psychology.	31
Psychology and Comparative Psychology	46

PART II.

Squirrels: Their Habits and Intelligence, with Special Reference to Feigning, with an Appendix	52
Hibernation and Allied States in Animals.	79

PART III.

The Psychic Development of Young Animals, and its Physical Correlation	113

THE FUNCTIONAL DEVELOPMENT OF THE CEREBRAL CORTEX IN DIFFERENT GROUPS OF ANIMALS

THE PSYCHIC DEVELOPMENT OF YOUNG ANIMALS AND ITS PHYSICAL (SOMATIC) CORRELATION, WITH SPECIAL REFERENCE TO THE BRAIN

PART IV.

DISCUSSIONS ON INSTINCT

ANIMAL INTELLIGENCE

PART I.

ANIMAL INTELLIGENCE, AND HOW TO STUDY IT.*

Is there any sane human being who is uninfluenced by the advent of springtime? Bright skies and balmy air have no doubt much to do with that heightened good-feeling which we experience, but do they explain the whole change in our being at that time? The budding trees and the springing grass have no small share in the happy effect, but even yet the analysis is far from complete. With no insects on the wing, no birds in the trees, no squirrels frisking among the boughs—would spring be spring? Are we not also influenced by the effect of the great contrasted change in our fellow-men?

The truth is, that no normal person is utterly indifferent to the world of life about him. But when squirrels frisk and birds carol, why is it that we regard them differently from mere mechanisms worked by a string? When we cage the bird or the squirrel, and they become tame, why do we apply such terms as, "dear little fellow," "sweet pet," etc? Why does the

---

* An Address delivered to pupils of the Bishop's College School, Lennoxville, October 1896.

fond mother thus address her infant? If the reply were—Because she cannot help it, would it not express an uncontrovertible truth? If you choose, I will put the same in another form and say, because it is natural for her to do so. But these endearing terms are used by the father, the brother, the sister, and even the one who drops in as a casual visitor, and is not in the remotest way allied by blood. While in these instances there may be something in the intellect and feelings of each, not in those of the other, there is much in common—at least such we must assume, if we are to furnish any reasonable explanation of the facts. So that when the mother, the father, the brother, or the sister speak of the squirrel as a "dear little fellow," or the cage bird as a "sweet pet," they must recognise in him some of those same qualities which render the infant human being attractive. The fact is, we make the world of animal life about us a reflection of ourselves; we spontaneously implant in the bird and the squirrel qualities that are our own. They interest in proportion as they seem to embody the same thoughts and feelings as ourselves. We assume that they have the same pleasures, pains—even the same hopes and fears. Man makes himself the measure of all things when he follows primitive impulses, and this is the real explanation of the interest that the great mass of human beings take in the world of animal life in general, and more especially in these animals with which we are brought into daily contact.

It must be clear that this is perfectly natural, as much so as to feel a sympathy with our fellow-men, though in a less complete and perfect way. It is only when certain animals are believed to be disagreeable or dangerous that they become repulsive, and in proportion as they seem to approach our ways of viewing things and share our feelings, are we drawn to them.

If Tom Jones is not interested equally in all the dogs he meets, it is because he believes they do not possess the qualities of his own Carlo. They do not fit into his mental world as well, they are not equally a reflex of himself. Every boy if left to follow his natural tendencies must believe that his dog's thoughts and feelings are the counterpart of his own. Were it otherwise, why should he talk to his dog, play with him, impute motives to him, blame or praise him, etc. ?

The change of view so general among civilised people calls for an explanation : and here we must distinguish between the clearly defined belief of the philosopher and the loose views—probably little more than prejudices—of the vast proportion of people.

The creed of the many in Europe and America as regards the relations of man to the world by which he is surrounded is in no small measure shaped by religious teachers. The result of this has been that man has been placed on a pedestal—raised high above all other creatures—and that is quite right too, but with this there has been interwoven the idea of the great inferiority of the brutes—which is again correct—but then there were associated with these views others which, in my opinion, have served to divorce man's sympathies from his fellows lower in the scale, and to lead him to view them in a distorted fashion. In attempting to glorify man, many well-meaning teachers have thought that they must depreciate his fellows, even to the point of denying to the lower animals any intellectual life proper at all—all was to be explained by "blind instinct"; so that by the time Tom Jones became a mature man he was unwilling to believe that his dog thought and felt as he did, and especially was he disinclined to set forth any such creed in words, and by no means would he have dared to do so before his teachers, lest he should seem to be thereby bringing himself into comparison with a mere dog. "Is thy servant a dog, that he should do this thing?"

But Nature is often wiser than her expounders, and I venture to assert that it is, in spite of the indignant protest of some people, because we are in not a little so like dogs that a large part of our life is what it is, and I hope before I finish this lecture to be able to convince you, or, at all events, to indicate the methods by which you may convince yourselves, that Tom Jones was quite correct when he believed that his dog's way of thinking, feeling, acting, and being acted upon, are very much like his own.

We all find some people hard to understand, and in proportion as that is the case are such persons estranged from us, and this is inevitable for the reasons that I am trying to set forth, viz.: that community of thought and feeling is essential to beget sympathy, interest, etc., and when they exist, and in proportion often as they are found, do they bind people together. You see these principles ilustrated in every school. John takes to James because, perhaps, they were at some other school together, and found they had common views and interests, and John finds it hard to get into sympathy with Harry, because they seem to have so little in common. John likes cricket and Harry lacrosse; John prefers to go a-fishing, and Harry to race across country; John is fond of quiet fun, and Harry of the boisterous, rollicking sort of amusement, and so one might proceed to illustrate at great length. Now and then, however, we meet a person who seems with extraordinary ease to be able to enter into the frame of mind and feeling of a large number of persons. Such people have, we say, wide sympathies, and when rendered intelligent by education they take broad views of things, and if possessed of vigorous intellect and strong will, they are likely to become leaders in the community. Possibly they may become distinguished in literature or art, because of that power they have of feeling themselves into many situations of interest to their fellow-men. Now I take it that Shakespeare was a man who possessed this faculty—though in an eminent degree combined with many others, as, for example, power of visual imagination and word-painting. But he might have had the latter and a host of other powers developed in the highest degree, and yet not have been a Shakespeare. He could not have felt like all the different characters which he put into even one of his marvellous dramas. In the course of one's limited experience he will meet people who have this power to put themselves in others' places, who are by no means Shakespeares, but who, notwithstanding, have in this invaluable endowment an affinity with the great poet; and you will generally find that such people are kind, slow to condemn, moderate in their censure and just in their estimates, all of which is more or less dependent on their ability to put themselves in the place of others—many others of different psychic make-up. Such are the people too, who are best adapted to understand animals, although they may, or may not be able to explain their mental qualities to others. There is such a thing as feeling one's way to truth when hard logic and cold philosophy are unavailing. You will, of course, not mistake my meaning here. I do not advocate the substitution of sentiment for cool, many-sided deliberation, but the putting one's self in the place of our fellow-men and the lower animals, and thus attempting to understand them. Indeed I would go so far as to say that this is the only way to make any real progress; all other methods must be aids to this final attitude of the mind.

In the understanding of the lower animals we must each become as a little child, and I know of nothing in which this is more literally true than in the study of comparative psychology—the psychic nature of animals lower than man.

To illustrate my position further: If you could have traced from early infancy up to the present time that one of your school-mates with whom you are most intimately acquainted — noting carefully every tendency, every important change in his circumstances and in his development, would you not be in a vastly better position to understand him than you can be at present? Perhaps you sometimes feel that you are incorrectly judged by others, and that if they only knew your past they would think differently. Indeed, what human being would not stand before us in a different light if we knew his whole history—and this resolves itself into his mental history finally. And then how incomplete is this until we go back beyond the individual, and look into the history and qualities of his ancestors, for after all, we are very much what our ancestors have made us; in other words, past history is determining in no small degree present events—the thoughts and feelings of our ancestors had, to say the least, no small share in moulding our own mental life. If one observes closely he will find that the resemblance to parents is just as close mentally as physically. I must not, however, dwell now on a subject so large and so important as heredity.

But if it is essential that we know the history of an individual human being to really understand him, it is almost, if not quite as important in the case of the dog or other of our dumb animals. Most of the dogs now in my own kennel were born and raised therein. Did time permit, I could interest you, I think, by showing how certain traits of these animals, which contribute to give them their individuality, are to be accounted for either by incidents in their history or by peculiarities which showed themselves soon after birth, and which were in all probability inborn. To illustrate by a single case. A collie of mine is very shy, so much so that an observer recently remarked that he seemed to act as if he were afraid of being beaten. As a matter of fact that dog has always been treated with the greatest consideration in view of his infirmity. He has improved considerably, and the causes of this improvement in his case I well understand. For the primary cause of the shyness in this case one must make a study of his ancestors.

Not only is it necessary in order to understand the individual dog to begin with him at his birth and to follow his history throughout, but such a course is essential for the comprehension of the nature of dogs in general, and, personally, I am deeply convinced of the importance of such investigations, after having been engaged in them for some years.

When such studies are carried out on representatives of different groups of animals, and on different breeds or individuals, one's conceptions of the true nature of animal intelligence—or, to use a more comprehensive term, the psychic life of animals—is vastly widened and altogether more correct in every respect.

In studying together, for example, a litter of puppies and a litter of kittens, the lines of development are found to be almost parallel for a time, then to diverge more and more. The same applies to the individuals of the litter, even though the circumstances under which they are reared are the same.

By this method of comparative study, questions as to what is common to the race and to different races closely allied, the relative strength of the individuality of members of the same litter or family, the influence of the surroundings, including in this all that we mean by education, and a whole host of other problems arise, and are to some extent solved. By varying the conditions under which the different members of a litter are reared, we may gauge the effect of the influence of the environment on the members of the little animal community. There are few more interesting problems than the relative power of animal tendency and of environment. The question as to what John Brown may become as the result of education, knowing the nature, and to some extent the strength of the qualities that were born in him, is of vital moment. But such a problem can be far more readily worked out for a dog than for a human being, because the nature of the dog and of its whole environment is simpler. Of late years much attention has been given to the study of the development of the infant from birth onward, and few psychologists would now doubt that the science of the mind has been put upon a sounder basis in consequence.

It must be apparent that such investigations are of the highest importance to all those interested in education. If we are ever to arrive at scientific, and therefore sound methods of education, it must be by a study of the true nature of the mind of man, and surely this will be advanced by a careful investigation of the psychic life of simpler natures, that is to say, of the lower animals. Of course the higher in the scale the animal studied, the nearer we are approaching on the whole to man. I say upon the whole, for it does not follow that in all respects the monkey, for example, a creature of superior intelligence to the dog, is more like man. In docility, some dogs at all events, are far in advance of monkeys, and in this respect nearer to man. A monkey is often a most perverse creature, even when plainly possessed of considerable intelligence. Nevertheless, there is no denying that a large ape approaches the psychic *status* of man more nearly than the most intelligent dog.

There are comparatively few people of intelligence in these days who would explain everything in the mental life of animals by instinct. But among those who make a special study of mind—the psychologists—and perhaps the biologists might be added, there is considerable diversity of opinion as to the exact nature of animal intelligence. One very distinguished writer would deny the power of thought proper to any creature that did not use language—articulate words. He would even go so far as to affirm that man himself can only think in words. But plainly his definition of "thinking" must be very restricted; it must be confined to a very few mental processes, and leave out a vast amount of what enters into the daily mental being of every man. There are others that would not go so far as this writer, who, nevertheless, deny to animals the power to perceive relations and to reason. When a dog appears to act as if he had reasoned, those who hold such views would explain by admitting that the animal had profited by experience; they would concede that he was intelligent, but claim that his apparently rational action was merely the outcome of mental association, or a use of "sense-experience." When, for example, a dog or a cat opens a door by manipulating the latch, writers of this school deny reasoning or any analogous processes, but explain the action by utilisation of sense-experience under the law of association. The dog somehow on one occasion, more or less accidentally, opened the door by using his paws or teeth on the latch, and this at once established an association in sense-experience; hence any future repetitions have nothing to do with any process of reasoning to the effect that if the paw be used on the latch the door will open. On the contrary, such writers deny the power to the animal to perceive any such relations.

This theory reduces the mental life of the animal very considerably, and restricts the dog's thoughts within a narrow compass. But is there not a danger of cutting down the possibilities of animal intelligence too much, and of assuming that in the mental life of the great mass of mankind there enter more of those higher intellectual processes (conceivable and possible at times) than there really are even among civilised men, not to mention the savage at all? We hardly realise, I fear, in how narrow a groove many minds move for the greater part of the waking period of every day; and this will hold, whether we take the case of those whose lives are one monotonous grind, or those who limit their thinking by devotion to some one simple but absorbing pursuit.

Let me illustrate by the case of a student who is passionately devoted to cycling. I know of one such case. His father is a professor, and in speaking of his son's absorption in this subject to the neglect of his studies, he expressed himself somewhat thus: "I would not mind if my son spent a couple of hours a day on the bicycle, and would forget it for the rest of the time; but wherever he is, he seems to think of almost nothing else, hence he cannot study successfully." Probably you can call up pretty well the condition of the mind of this youth. He sees bicycles, he feels bicycles, he beholds race-tracks and crowds, he hears applause, he receives prizes in imagination, etc., etc., and this over and over again with little variation. You might construct a diagram giving a representation of the probable thought relations in his case, seeing he lives in a realm of "sense-experience," one in which the perception of relations only occasionally enters, if we are by this to mean such perception as is impossible to the dog.

Indeed, how much is there in such mental states that is impossible to the dog? If for cycling we substitute hunting, the case will be clearer. Do you think that the pictures of the hunting-field on which the man feasts, differ much from those the dog calls up?

It is quite true that the man *can* think otherwise about his hunting, cycling, etc. He can discuss with his fellows the causes of successes and failures; he can even write a treatise on hunting. But when the hunter thinks matters over in his own mind after a day's sport—when he goes so far as to seek for explanations, when he must perceive relations, do you conceive that his state of mind is absolutely impossible to the dog—that is, most of it? The case for the ape is stronger still, when we consider the wonderful resource he sometimes shows in protecting his offspring, in co-operative defence, etc.

Some writers, who assume a very conservative position in regard to animal intelligence, lay down the rule that we must in no case interpret an action as the outcome of the exercise of a higher faculty, if it can be interpreted as the outcome of the exercise of one which stands lower in the psychological scale. Hence they think that many of the actions of such animals as dogs, cats, etc., which some would attribute to reasoning or an analogous psychic process should be explained in some simpler way. But why should we bind ourselves by a hard and fast rule like this one? Is it not the truth at which we wish to get? For myself, I am becoming more and more sceptical as to the validity of simple explanations for the manifestation of animal life whether physical or psychical. It is true the whole matter is made easier for the student, in that he can the more readily grasp and remember the opinions of others, processes which take up a large part of his time. If we analyse our own actions, especially the motives for them, with special care, how often do we realise that our fellow-creatures have judged us hastily and imperfectly. That our motives for actions are often found to be far more complex than even we ourselves supposed at first, is a view of the case that was presented to me when a youth by a very profound and much-esteemed school-teacher of mine, who was good enough to give me occasionally the benefit of his own thinkings, and the more I examine myself, and look into the psychic life of others, the more do I feel the force and justness of my teacher's view.

Are we not in like manner too ready to adopt simple —unduly simple—explanations of the actions of the animals by which we are surrounded? You will, of course, not suppose that I would claim that the motives —using that term in the widest sense—which actuate them are of equal complexity with those that determine the actions of men in many cases; but in all discussions on animal intelligence, and the entire psychic life of creatures that are, on the whole, lower in the scale than ourselves, we must be careful to distinguish difference in quality from difference in degree. And in the investigation of so important a distinction it seems to me of the greatest moment to compare the human being at various stages of his development with the lower animals in a corresponding way hence, the, to my mind, absolute necessity of investigation of the psychic development of both the lower animals and of man. A dog at different periods of his existence stands, as it were, on different psychic planes. He leaves some features of his early life behind him for good—not many, however, while he adds and adds new developments which, in different dogs, vary with their special experience, but not enough to obliterate the general characteristics of the canine mind. Just the same may be said of the human intellect, and there are few more suggestive or fruitful studies for those who have an interest in such investigation than the comparison of the child and the dog at their different epochs of development. Of course, the parallelism is clearer during the earlier epochs, and the dog runs through the main stages of his psychic life very much more rapidly than the child; but apart from the use of language and the special peculiarities of the psychic activity dependent on this, there is a closer resemblance—at all events, if we restrict our comparisons to unlettered, and especially uncivilized men—than most persons would suspect, or, owing to prejudices, would be inclined to admit. Nor would I confine this statement to the dog, for a study of a kitten for 135 days, from birth onwards, was a revelation to myself, though I had been a steady observer of animals for a long period of years. The amazing persistence and intellectual resource shown by this kitten were such as to remind me of nothing more than the conduct of a child of unusual determination and intelligence—in fact, just the sort of child that I should expect to succeed in the world, no matter what the obstacles in its path.

Nearly ten years ago, in a paper published in the *Popular Science Monthly*,* I made the statement that "Many of the performances of the lower animals, if accomplished by men, would be regarded as indications of the possession of marvellous genius," and I see no reason now to change that opinion.

That man can lay out the line of a railroad through the trackless forest, over lofty mountains and across deep valleys, is indeed evidence of wonderful mental achievement. But if the surveyor could dispense with all his instruments and mathematical calculations, and were in possession of some mental endowment by which he could straightway indicate the correct path, would his performance not be immeasurably more wonderful? And would we attempt to belittle it by assuming that it did not involve reasoning and the use of syllogisms. If genius has any one quality about

---

* March 1887.

which men are agreed, it is that its performances are inexplicable, either to the individual himself or to others, in at all events, the most remarkable cases. Take as instances in very different lines of thought Newton's perceptions of quantity and space relations, and Mozart's of tone relations. These perceptions were immediate, and surmounted all ordinary rules of mental movement. But when a homing pigeon covers 500 miles in so short a time that the rate of speed rises to 40 or 50 miles an hour, showing how straight is the path by which it reaches its home, we are ready to class the performance as wonderful for a bird, but not on a par with any feat of human genius. So far as I know, no one has as yet explained such a performance. I have studied this subject, and made some experiments with homing pigeons, but whether we explain the matter as the exercise of very accurate perceptions of landmarks—which is not an explanation without great difficulties when long distances are involved—or whether we give up the problem and say we have no experience which enables us to understand it—the result is still marvellous, and is closer to the performance, of genius than anything else to which it can be compared. It is to be remembered, too, that we may find even in imbeciles or idiots certain psychic capabilities as, *e.g.* for music, developed to an amazing extent, so that the generally low intelligence of a pigeon is not to be set up as a plea of belittlement of its homing performance.

Therefore, while man is a law unto himself, and to a certain extent a law to all other creatures, while he must look within to understand himself and use introspection in attempting to get at the nature of the psychic life of the lower animals, he must also recognise the limitations of this final method, and realise that he may stumble on problems regarding both himself (in the case of the genius) and the lower animals which are insusceptible of solution.

I have always thought that this gratuitous assumption of inferiority in *all* respects of the lower animals was an evidence not only of man's unbounded conceit, but is further evidence that he had not even realised the nature of the problems to be solved.

The more I study the subject myself, the more do I hesitate to adopt outright the explanations already given by those who have written on the subject.

I think we have of late made rapid progress, but there is still great need of observation and experiment without bias. All may gain in modesty and in knowledge who will in the right way study animal intelligence. Few people have the qualifications of long and intimate association with animals, by habit of personal introspection, etc., etc., to work out the deeper problems, but we may hope that the number will increase rapidly in the years to come.

To sum up then, somewhat imperfectly, it would appear that all human beings, whether civilised or savage, naturally have an interest in animals because, consciously or unconsciously, they assume that they resemble themselves psychically. Possibly the fact that savages seem the better to understand animals in some respects is owing to their being able the more readily to place themselves on their psychic plane. Closet comparative psychology cannot hope to accomplish much. He who would understand animals thoroughly must live among them, endeavour to think as they think, and feel as they feel, and this at every stage of their development. Observation, experiment and introspection are all essential to the student of comparative psychology, but we must recognise that there may be problems in both human and comparative psychology that so far, at all events, as certain individuals are concerned, are possibly beyond solution. Such are probably few in number, and under attack by new methods may be rendered still fewer; nevertheless, it is healthful for man to say now and then after the fullest study—" I do not know." There are states of our minds which no doubt bear a closer approximation to those of animals than others, and these should be seized upon and analysed, if we would understand the mental life of animals. No small part of our psychic life differs from that of animals rather in degree than in kind. Nothing is to be gained for any cause, however, by overstating the case, and it is a mistake to claim that between the highest men and the most intelligent animals there is not a vast difference, even if we do not go so far as to say there is a great gulf fixed, as some appear to believe. This is another thing, however, from assuming that the same holds for the most lowly developed men and the most highly developed animals. As to the differences in the latter case, there is room for great diversity of opinion in the present state of our knowledge.

Experiments with the lowest classes of men and on all kinds of animals are urgently needed.

In the meantime a modest, enquiring, open state of mind is that most becoming and helpful.

COMPARATIVE PSYCHOLOGY: ITS OBJECTS AND PROBLEMS.*

THE term comparative psychology, in its modern sense, gives us the widest desirable scope as including all that pertains to the mind or soul of the animal kingdom. It may have been at one time considered as highly impertinent to ask whether the lower animals possess mind, and to substitute the term soul would have been dangerously suggestive of heterodoxy of a type rapidly to be extinguished. However, few persons of any degree of culture will now be found prepared to deny that the inferior animals have minds. The questions now to be settled are: What kind of minds? In how far do they resemble, and in how far differ from, our own? Few, it is true, have considered that they sufficiently resemble the human mind to make it worth while to investigate the subject at all. Probably the great mass of persons have been led to believe that man does and always has occupied a distinctive and wholly isolated position in the universe of life—a centre around whom and for whom all other forms exist. This view seems to me totally unwarranted by the state of our scientific knowledge at the present day. Further, it is a view not only without scientific foundation, but calculated to lead to pernicious practical results.

By experiments on the lower animals, and by this means almost wholly, has the science of physiology been built up. We argue from the case in animals to the case in man, and consider the inferences thus derived valuable, even final—possibly too much so;

---

* A Presidential Address delivered before the Society for the Study of Comparative Psychology in 1887.

but we are apt to ignore the psychological similarity. From experiments on the brains of the lower animals we argue as to the nature of the brain of man. Why not pursue the comparative method for the soul?

This condition of things can be traced to the influence of views still surviving, unscientific, as we believe, as to man's origin and place in the universe. At all events, such views exist and influence practically our treatment of the lower animals. Where man is concerned, their rights are very seldom considered. The question is not raised as to whose rights are paramount, but it is tacitly assumed that when man is involved the brutes have none. That such views have been up to the present time operative to the neglect, and often the positive annoyance, if not the actual persecution and death of unoffending creatures, will be perfectly plain to any one who will take the pains to examine into the case.

If there is to be order in the universe, it must be conceded that where respective interests clash in certain cases, that interest and that creature of less importance must give way to the one of greater importance; but man can never act righteously to his fellow-creatures lower in the animal scale till he recognises that he is of them not only in his body but in his mind; in other words, that they are truly fellows, or, as some one has expressed it, "poor relations." But let this not be said in any pitying sense, for it can be most clearly shown that in not a few respects not only are these "poor relations" equal but superior to man.

Physiologists have long been familiar with the higher development of the senses in animals below man. There is not a single sense that man possesses in which he is not excelled by some one animal, often immeasurably.

Many of the performances of the lower animals, if

accomplished by men, would be regarded as indications of the possession of marvellous genius. In the brutes they are regarded as the outcome of "mere instinct," by which is meant an endowment acting blindly and incapable either of philosophic explanation or of modification. While the fact seems to be that instincts, as they exist, are the result of inherited experiences accumulated through considerable periods of time; that they may be modified, and are constantly being modified by new experiences; that they may be lost or replaced; and much more that we have still to learn, many of the instincts of animals are so far removed from any knowledge or faculty we possess, that they are at present inexplicable. But man must learn to say "I don't know" about a great many things still, instead of assuming the validity of explanations which are not true solutions at all, but mere assumptions.

And at this point allow me to indicate a danger that should make us cautious and modest in attempting to explain the behaviour of animals. We infer from our fellowman's behaviour similarity of motive and mental processes to our own under like circumstances. We find, the more experience we have, that we are often at fault as to both. And when we are more free from the thraldom of so-called systems and methods in education, we may learn that the activities of the human mind can not be reduced in all persons to precisely the one plan, like so much clockwork. This may mar somewhat the completeness and beauty of our philosophy of education, but it may also in the end conduce to human progress by providing the greater freedom, and end in insuring an individuality of character which seems to be now rapidly disappearing. Now, if individual men so differ in psychic behaviour, how much more is it likely that still greater differences hold for the lower animals! An objection may be based, however, on this, to the whole study of comparative psychology. The objection holds to some extent even for human psychology; but, as we infer similarity of behaviour in men to denote similarity of inner processes, so are we justified in the same as regards the lower animals, though it must be conceded somewhat less so. We must always be prepared to admit that there may be psychic paths unknown and possibly unknowable to us in the realm of their inner life. But if we regard man as the outcome of development through lower forms, according to variation with natural selection—in a word, if a man is the final link in a long chain binding the whole animal creation together, we have the greater reason for inferring that comparative psychology and human psychology have common roots. We must, in fact, believe in a mental or psychic evolution as well as in a physical (morphological) one.

It is not inconceivable that special faculties which do not exist in the lower animals have been implanted in man; but the trend of investigation thus far goes to show that at least the germ of every human faculty does exist in some species of animal. Nor does such a view at all derogate from the dignity of superior man, while it links the animal creation together in a way that no other can. It opens up the subject for genuine scientific study; it tends to beget a respect for the lower creation, which, while it fosters modesty in man, also furnishes a foundation for broader sympathy with those lower in the scale. The opposite view may lead to our pitying the brute, but can scarcely yield as good moral fruit. Let but an individual man assume that, by virtue of something he possesses, he is radically different from his fellows, and what is the result? Your genuine aristocrat (in feeling) is a sad stranger to humanity in general.

But where shall we draw the line? Formerly the line was drawn at reason. It was said the brutes can not reason. Only persons who do not themselves reason about the subject with the facts before them can any longer occupy such a position. The evidence of reasoning power is overwhelming for the upper ranks of animals, and yearly the downward limits are being extended the more the inferior tribes are studied. Perhaps the highest faculty man possesses is that by which he generalises and forms conceptions of the *abstract*. That animals have imagination or the power to frame mental pictures of absent objects the grief of the dog at the absence or loss of his master amply proves, as does also the capacity of animals to dream. If, as some assume, abstraction is a necessary part of reasoning, then it must of course be conceded that animals have the power of framing abstract conceptions. There is a certain amount of evidence that some animals can count within narrow limits. It is scarcely possible to account for the conduct of the horse, dog, elephant, and ape, under certain circumstances, without believing that they have the power to generalise upon details. Once concede the power to form abstract ideas, and there is then the basis for any other faculty man possesses that is considered usually as peculiarly his.

Have animals a moral nature, or are they capable of forming a conception of right and wrong? The answer to this introduces the question as to method of comparison. Should the highest of the inferior animals be compared with the most civilised races of men, or with man in his most degraded condition? That neither of these comparisons is just can be shown. As capacity for education is one of the best evidences of mental ability in both man and inferior animals, and as man's civilisation is the outcome of his own intellect, he must be credited with this as evidence of his superiority.

It is to be remembered, however, that each marked advance in progress has been made by the few great intellects that have appeared, and only accepted, not originated, by the many; that but for permanent records in language, much of man's civilisation would have been lost as rapidly as acquired; that man's civilisation is the growth of thousands of years, beginning with a condition of things scarcely if at all higher than that now known to some tribes of animals; that what any child becomes is really largely dependent upon the training it receives; the child of the savage, and that of the civilised man, can not be compared any more than the latter and the inferior animals. Now, the reverse of all this holds for the lower animals. So far as any systematic training from man is concerned, they are very much as they were thousands of years ago. Before it were possible absolutely to compare the highest man and the highest animal, it would be necessary that for ages the effect of culture should be tried on the lower animals. The astonishing results achieved in the lifetime of a single animal, and the results attained by the creation of hereditary specialists as among dogs, put the whole matter in a light that shows our usual comparisons to be somewhat unfair. If the highest among dogs, apes, and elephants be compared with the lowest among savage tribes, the balance, whether mental or moral, will not be very largely in man's favour—indeed, in many cases, the reverse.

We are not contending for the equality of man and the rest of the animal kingdom; even assuming that the child and the dog have equal advantages, the child will still be in many respects superior to the dog; but we are desirous of pointing out how much has been overlooked in all these comparisons between man and the lower animals. It will be noticed, that all those species of animals, which have for ages been in contact with man, have made great advances over their wild progenitors, evidencing a capacity for education—mental and moral—which is one of the best demonstrations of superiority.

The assumption that man is only accidentally the superior of the brute would but lead to confusion, for it must be admitted that there is a scale, and that man ranks first. We are simply desirous of doing the lower creation that justice which we feel assured has not yet been allowed them, and of seeing the human family interested in those that we think scientific investigation is proving constantly are much more our fellow-creatures than has generally been supposed.

If we compare the intelligence and general rectitude of behaviour of our best races of dogs with the same in any of their wild carnivorous allies, we are astonished at the great difference in favour of the dog. To what is this due? Largely to what he has become by virtue of association with man for hundreds if not thousands of years—that is, to education, after a fashion. Nor is such influence confined to the dog. Any observing person, of moderate experience in travel, can call to mind numerous instances of members of different classes of animals trained to the performance of many feats demanding intelligence. But while, in an irregular way, dogs have been trained to certain duties for the benefit of man for a considerable period, it can not be said that any one of the tribes of the lower animals has ever been subjected to any such mental or moral discipline as man receives and has received for long ages. We have ample evidence, in the condition not only of savage man, but in the neglected classes of large cities, as to what man would be without such culture. Sufficient has been said, it is believed, to show that we are not yet in possession of enough facts to enable us to determine exactly the limit of mental and moral capacity in the lower animals. As yet we neither know adequately what they are or of what they are capable. Both these subjects are worthy of human investigation. Their elucidation must tend to give man a better knowledge of himself, if only by contrast.

To return to the question of the moral nature of animals. The study of the dog alone, in the light of observations accumulated in the literature which are often true of special individuals in a degree not of the average animal (a fact which does not, however, at all invalidate their force), or the study of any dog we may ourselves own, can not but convince us that a sense of right and wrong is possessed by that animal. It may be that the dog does not rise to these conceptions as understood by the learned divine discoursing from the pulpit; but neither does a large proportion of the congregation when transacting the business of the week. It may be, and perhaps is, largely true that the right with the dog means what is in accord with his master's will; that is, the dog may end at the stage in which every child, even the most highly endowed, is found at *some* period of his development. It is a condition unquestionably in advance, by far, of that of scores of tribes. Moreover, as in the child, and the less endowed morally of men, even such ideas of the right are powerfully operative in producing courses of useful conduct. They lead to action on the one hand, and to restraint on the other, instances of which, in the case of the dog, are abundant, and some of them of a most touching, we might almost say ennobling, character. To affirm that the idea of right and wrong of the lower animals does not rise above the hope of reward and the fear of punishment is not to keep to the facts, unless we include as the only reward, in many cases, the master's approbation, and the only punishment his displeasure. When a child arrives at such a stage of feeling, most persons would not be inclined to deny it a moral nature and a very good one, too. We might almost speak of a dog having a religion, with man as his deity. But as a whole host of qualities— some of them difficult to classify — go to make up the character of the human individual so developed and balanced as to deserve the epithet "gentleman," so there are many qualities in the best specimens of the canine race that we can practically appreciate better than define.

In all such discussions it must be borne in mind that if we adopt the theory of organic evolution, we are almost bound, of necessity, to a belief in the origin and gradual development of mind from the faintest glimmerings of consciousness, in the simplest protoplasmic creatures; and that system will be most philosophical and complete which can fill up the gaps between the lowest manifestation of any quality and the highest. Hence, many are inclined to believe that the great distinction between man's faculties and those of animals lower in the scale is *difference in degree and not in kind*, certainly in so far as they run parallel. Such a view does not prevent our conceiving of additional forms of psychic activity not represented in man as the possession of the brutes. That such seems probable will appear when we discuss some of the problems still demanding solution. Nor does such a view imply that there may not be avenues of knowledge of a special kind open to man which are closed to those lower in the scale, such as a special revelation from a higher source. So far as we see, indeed, there are no theological difficulties any more than with evolution as ordinarily applied to animal and plant forms.

Man's present superiority over the lower animals is

traceable in large part to his eminently social tendencies, resulting in the division of labour, with its consequent development of special aptitudes and its outcome in the enormous amount of force which he can, on occasion, bring to bear against the various tendencies making for his destruction. Indeed, the isolated individual man is scarcely as well prepared in the struggle for existence as most other animals. But the extent to which animals do continue, it may be in pairs or in larger numbers, to defend themselves against enemies; hunt down prey; rear young; elude enemies; overcome difficulties in travel; work in concert in the preparation of dwellings, and in many other instances, has been but inadequately considered. And in many such cases it is quite impossible to explain these things by that refuge of the unthinking or prejudiced, "instinct." The limits of an address of this kind do not, of course, permit of detailed evidence being adduced for the views maintained. Such evidence is, however, within the observation of all to some extent, and is, so far as the literature is concerned, found in elaborate form in the admirable writings of Romanes and Lindsay more especially. Thus much by way of clearing the ground, of preparing the mind for a careful and earnest study of our fellow-creatures of the lower grades, without prejudice, and without fear of any loss of self-respect by the concessions we may be obliged to make.

As to how, so far as the study of comparative psychology itself is concerned, the objects of this society may be best advanced, let me now endeavour to indicate briefly. A great part of the material available is found in literature of very varying reliability. In many cases there is so obvious a prejudice in favour of the particular animals whose performances are described, that very large deductions must be made. We shall do well to be more than cautious in what we accept. At the same time, much that can not be regarded as wholly reliable may prove suggestive and serve as the starting-point of investigations. But there is no reason why many points now bearing the character of uncertainty and indefiniteness might not be submitted to the test of experiment. Doubtless not a few supposed facts would vanish into thin air if subjected to such examination. However, I must at the same time state that a careful perusal of the accounts of the experiments of even the most skilful investigators by this method, with its clearly defined but artificially arranged conditions, has convinced me that such do not wholly meet the case. They bear with them the danger of fallacy against which one must constantly be on the watch. It must always be considered that the great question is, not how an animal's mind *may* act, valuable as that may be, but how it normally *does* act; that is to say, what are the natural psychic processes of the class of animals under investigation? The same caution, in drawing conclusions, must be observed in the allied science of physiology, one in which the conditions can be much more accurately regulated. Plainly, it will be desirable to keep our *facts* very sharply apart from our explanations. The science of psychology is a very youthful one, that of comparative psychology still more so; and, at the present stage of the science, any one who contributes a single fact will be a real friend to their progress. We must endeavour to secure a large number of correspondents who will furnish accurate accounts of phenomena in this realm, of which they have been themselves the observers. We must place all material coming at second-hand by itself, not as worthless, but as calling for special scrutiny. But so long as we have facts only, we have no science; such, indeed, are as the wood and stone for the building, and, unless worked up into scientific form, may prove an incumbrance. Let me, then, briefly indicate some of the problems that have seemed to myself and others as most urgently demanding solution.

One of the questions still far from clear is that which we had under discussion last year, viz.: In how far can the lower animals understand man's various forms of expression, especially his spoken words? *A priori*, we should not expect that creatures unable to invent words should have the capacity to understand them in the sense in which man himself does. I am inclined to think that more has been claimed for the inferior races of animals in this direction than an exact examination of the subject will warrant. On the other hand, we have probably very much underrated their capacity to comprehend our various forms of unspoken longuage. The subject calls for close observation. A kindred problem is the degree to which various kinds of animals can communicate with one another. This is a much more difficult subject, and it may prove that the creatures we despise as so very much inferior may have modes of subtle communication which we are, possibly, incapable even of comprehending.

The whole subject of the senses of the lower animals is a field for investigation both by the psychologist and the physiologist; all the more important, as it is scarcely possible to understand one form or degree of sensation adequately, except by comparison with its lower and higher forms. The field is as yet but little tilled, but enough has been done to suggest this very important question: Do the senses of the lower animals and those of man differ only in degree, or also in kind? Is the sense of smell, *e.g.* in the dog, merely more acute, or is it not also characteristically different? The latter seems the more probable, when we consider how different the hearing of man is in some respects (music) from that of other animals, even the dog.

Among wholly unsolved problems ranks the nature of the mental processes by which many different tribes of animals find their way back to the place from which they have been removed, when the distances involved are great, and often when they have never travelled so much as once the way by which they return.

Akin to this, possibly, though perhaps quite different, is the question as to the nature of the faculties by which animals are enabled to migrate. "How a small and tender bird, coming from Africa or Spain, after traversing the sea, finds the very same hedgerow in the middle of England, where it made its nest last season, is truly marvellous" (Darwin). We are much in need of more *facts* in regard to the migrations of animals; and it is hoped that the systematic work recently inaugurated by the American Ornithological Association may lead to useful results in this field. With regard to the so-called "homing instinct," it has been noticed that savage or semi-savage man possesses a power of finding his way in the trackless forest by more accurate observation than that of which the civilised man seems capable. While this throws light upon the case of the lower animals, it does but very inadequately explain it. It may turn out that both of these puzzles are susceptible of simple explanation; but at present they strike me as rather belonging to that class of psychic phenomena, the meaning of which can be but inadequately understood by man, owing to his not possessing the requisite faculties or those faculties in sufficiently powerful or acute development. The performances of a Shakespeare and Scott in literature, or a Beethoven in music, to the mass of men, must be but imperfectly understood in any proper sense of *realisation*. Probably these sons of genius could have given little account of the "manner of it" themselves. We might hesitate to call such faculties as the above in the lower animals genius, or to acknowledge any kinship; but genius among men is often as limited and as disassociated with general mental power as are certain marvellous faculties in the lower animals. It may be that migration is accomplished by means of some forms of acute sensation, according to which the animal acts more or less blindly. Plainly, no mere restless impulse can account for the performance, though it may initiate it. These and many other problems are before us; and, like most recondite problems, they will require the labours of many, each bringing his little for their solution. But is it not worth while? Man can not live by bread alone. We hunger for completeness in our knowledge and harmony in our philosophy. But, apart from this philosophical satisfaction, it cannot but prove for the interests both of man and the lower animals that the latter should be better understood.

Belonging, as most of you do, to the veterinary profession, or, as I should prefer to call it, the profession of comparative medicine, either as students or as practitioners and teachers, the more you comprehend the mental workings and modes of expression of your patients, the more successfully must you arrive at an accurate knowledge of their symptoms, and so be the better prepared to relieve the suffering among them, and in so doing also advance man's material interests. To you, at the present time, must we especially look for diffusing more enlightened and humane views, views worthy of this renowned School of Comparative Medicine, which many of you have come so far to attend. It will be for you to intervene in cases of public panic, like that witnessed in connection with the recent hydrophobia scare; reassure the public mind, and protect our fellow-creatures of the lower ranks from needless molestation. There is probably no class of men whose daily life-work gives them so large an opportunity for at the same time acquiring and diffusing truer views in regard to the lower animals. Your enthusiasm and success during the first year of our existence as a Society, have been a matter of equal surprise and delight to me, especially considering how fully you are occupied with the ordinary duties of your profession. We hope to enlist the interest of others and bring them into our ranks; to accumulate a library of books bearing on this subject; secure a large number of correspondents from widely separated parts of the continent, and in various other ways stimulate the study which we feel calls for and is worthy of man's earnest attention.*

I cannot close this address without making grateful reference on behalf of this Society to the kind manner in which, in many ways, Principal M'Eachran, and the Professors of the Veterinary College, have lent their support to our projects.

COMPARATIVE PSYCHOLOGY.†

IN entering upon the third year of our existence as a Society, it has seemed to me that it might be encouraging to the older members and instructive to those who are meeting with us for the first time, to review the work of the Society for the past two years; to point out what we have tried to accomplish and what has been actually achieved.

---

* This young Society, so far as known, the only one in America for the study of Comparative Psychology, is composed at present almost entirely of the students and teachers of the School of Comparative (Veterinary) Medicine in Montreal, though its membership is open to all eligible persons.

† Read before the Association for the Study of Comparative Psychology in connection with the Montreal Veterinary College, in 1888.

Believing that men who had chosen Comparative Medicine as a career, must have some real liking for those animals, at least, which are classed as domestic, if not for all creatures that breathe the breath of life, and feeling assured that a knowledge of the mental constitution of animals must prove invaluable to the veterinary surgeon in the diagnosis and treatment of the diseases of his speechless patients, in the latter part of the year 1885, I called together such of the students of the Montreal Veterinary College as were attending my own classes in physiology, and suggested the desirability of forming some sort of association for the attainment of these objects. Those addressed responded to my proposals as only young men can. Soon almost every student in the College joined us. The Principal and Professors aided, both by smoothing the way and by active and cordial co-operation. A spacious and comfortable room was kindly placed at our disposal in the Veterinary College building in which to hold our meetings. As the project was tentative, we did not think it well to fetter ourselves with many rules or regulations. However, on commencing our second year, we felt warranted in giving our Association a name, providing for it a constitution and bye-laws, and taking such other steps as tended to show that organisation was warranted as a natural result of growth and development.*

In order to present the history of our Association within a small space, you will bear in mind that the accounts of papers read, and the discussions ensuing, must appear in very condensed form; and that the comments I have now to make on them must be few, and rather indicative of the line of investigation we have

---

* Thus far the Principal of the Veterinary College, D. M'Eachran, has been the Honorary President; T. Wesley Mills, Professor of Physiology, M'Gill University, President; W. J. Torrance, Recording Secretary. The other offices have been filled by different members of the Association, including the Professors of the College.

followed, and should continue to pursue, than as statements of established results. Naturally, most of our studies, though by no means all, have been of the domestic animals, and, as was to be expected, the dog is the creature whose mental nature has been most frequently the subject of our enquiries—and this will likely be the case in the future, also, for many reasons; or if we can establish some conclusions regarding the psychic operations and development of any one of the lower animals, we then have more certain ground for comparison, even if we never succeed in showing that we have any warrant for interpreting the mental operations of inferior animals in terms of those of man. If we could establish a relative scale of intelligence for animals below man, much would have been accomplished. The first communication laid before the Society grew out of a paper read before the Veterinary Association by Principal M'Eachran. In this communication the behaviour of a dog that was manifestly possessed of unusual intelligence was described in detail. Among other evidences of this were his journeys to a baker's shop to purchase food for himself. Several such cases are on record, and as I shall have occasion to bring this matter before you again shortly, it will not be dwelt upon now. In all such cases we cannot be too cautious in the explanations we adopt. Mr Dawes, at the same meeting, sketched the history of a Cocker Spaniel that, in consequence of early training, would, on request, fetch any one of six different articles.

This case led to the important enquiry: In how far, or in what sense, do animals understand words? In the course of the discussion following, it was pointed out that dogs would answer to their names when uttered by strangers, in opposition to the view that the animal was guided chiefly, if not solely, by the general demeanour of the person calling the dog, the tone of voice, etc. The fact that each individual of a pack of hounds will respond to his own name is also significant. The observation that, as noticed by one member, his dog would answer sometimes to names very similar, as "Dick" and "Vick," was not without parallel in the case of men, and was explicable either by imperfect hearing or by inattention.

The case, as instanced by a terrier that seemed to hunt equally well for rats, whether "cows" or "rats" was the inciting term, did not furnish a wholly valid objection, it was thought, for in all such instances the accompaniments of the utterance of the mere word were of more significance than the word itself. I shall have evidence to present to you during this year which I think will make it clear that at least many dogs really do know their names in the same sense as very young children, if not even in a higher sense.

Frequently, during the past two years, the influence of breed, of the individuality of the owner or trainer of the animal, of food, and general treatment has been under discussion.

These questions are not only of the highest theoretical interest, but of the greatest practical importance. At one of our meetings certain members advanced a view favourable to a course of severity in dealing with certain horses, such, for example, as the "bucking" ponies of the prairies. The President believed that it was of the utmost importance that such a view should not be entertained by veterinary surgeons, and that efforts should be made to eradicate it from the public mind in so far as it really exists.

Most of the difficulty experienced in managing animals arises from their not understanding what is required of them, or from mental associations which have been established by previous unwise or cruel treatment. I cannot here refrain from stating the opinion of an eminent horse-trainer with whom I lately conversed. He holds that every horse should be broken and trained by some one more or less of an expert; that we expect too great a variety of performance from the same animal. Each is naturally, to a large extent, best adapted for some one kind of work—in a word, each is, to a large degree, fitted to be a specialist. But in this case a good many drivers would require to be " broken " also. The brutes are constantly suffering from the stupidity, as well as the moral obliquity of man, their controller, but not always and in all respects their superior. These remarks do not apply alone to the horse or the dog. All animals must first learn that they are to be subject to man when required; but, as I have always maintained, the highest results are to be secured only by kindness and discretion combined with firmness. A little reflection will show why this must be so. One does not facilitate the working of a steam-engine by any sort of forcible interference with the parts of the apparatus, but by supplying good fuel and duly oiling the machine where friction is greatest. So it is for man to study the mental machinery, so to speak, and provide those conditions most favourable to harmonious working; in a word, man must adapt to nature and not attempt to make nature adapt to his views. The latter he cannot do; her plan was laid before he appeared on the scene. If an animal is so stupid or so obstinate as not to yield to such treatment, then it should be abandoned, for it will not be worth any man's while to injure his own moral nature by what is really cruel treatment for the sake of the value of such an animal.

At another of our meetings Mr Miller referred to the case of a dog that was very anxious to accompany his master, resorting to the artifice of placing himself some two miles in advance on the road usually selected. There are many such instances, and it seems impossible to explain them, except by the exercise of reasoning power or some mental process closely analogous. But it must appear superfluous to contend any longer for the possession of such a faculty in the higher groups of animals at least. One of our members, Mr Metcalf, himself the owner of a large number of dogs, referred to the fact that one of them had a great dislike of beggars, tramps, and such like persons. From what I have read of similar and even more marked conduct, from much that I have seen, and especially in a young dog I now possess, I am almost persuaded that in certain dogs such hostility is inborn, and, in certain cases, hereditary. Mr Metcalf thought that the detention, without injury, of would-be thieves, as in a case he reported, was peculiar to the mastiff.

In February 1886, Mr John Miller read a paper on the dog, with special reference to the Scotch Collie, which brought out some interesting remarks from a member who had witnessed the training and performances of these animals in Scotland. Everything went to show that the collie dog is a specialist of marked aptitudes, the result of ages of training and selected breeding, though his general intelligence is also high.

At the following meeting Mr Ferron reported on the intelligence of a certain bitch he had observed. The animal imitated a cat in carrying kittens and in several other particulars; she was also remarkable in retentiveness of memory, and in other respects. This case was all the more valuable a study, inasmuch as the animal had received no training whatever.

The President instanced the case of a brindle bull-dog that had, on several occasions, found his way home, a distance of twenty-four miles, and in so brief a time as to indicate that he must have taken short cuts. Such cases suggest one of the most interesting and puzzling enquiries in the whole realm of Comparative

Psychology: the nature of the mental processes by which animals make their way back by a different route to places from which they have been taken. I have given the subject considerable attention, and I hope before very long to be able to throw some new light on this vexed question. At this meeting, a paper published by Dr Packard in the *American Naturalist*, for September 1885, on the "Origin of the American Varieties of the Dog," was read, on account of the great interest of the subject. When we consider how widely the dog has departed from all his supposed ancestors in his physical traits, we are amazed at the extent to which lower minds can be modified—we might almost say radically changed—by contact with the dominant mind of man.

This being the last meeting of the session, the President proposed certain subjects for study during the summer. These were put in the following form:

To what extent have the lower animals imagination? What animals dream? The persistence and modification of instinct. Is there a "homing instinct" or a "sense of direction" peculiar to animals? What groups of animals understand mechanical contrivances, and which can use tools? How far do the minds of animals become modified by contact with man? Have any animals a special aptitude or a peculiar faculty for determining where water is to be found? The special senses of the lower animals compared with those of man; feigning, catalepsy, etc., in the lower animals; a moral sense in animals below man.

It is not to be understood that our attention was devoted exclusively to the dog during our first year of existence as a Society, but it has appeared to me best to give a sketch of our investigation of each animal separately, so I now continue the account of our study of the dog during last year. Before doing so, it may be mentioned that the session was opened by the election of officers and the delivery of an address by the President, in which many of the topics proposed for special study at the close of the previous year were reverted to, and the objects of the Society indicated. As this address has been published,* and copies of it are already in the hands of most of the older members, I shall not further refer to it than to say that the subject has attracted attention generally, and its treatment, as was expected, has received some criticism.†

To return to our friend the dog. Early in the session, Mr Simpson made an important communication, the result of a careful study of a blind Pomeranian dog. He had proved conclusively that this animal understood his name, and also many other words, such as "sneeze," "bark," etc. The dog had been blind for two years, but had so made use of his remaining senses, and his mental faculties generally, that he seemed, except in special cases, but little worse off than before. He recollected well the location of stable objects, and was able to make his way successfully through the business portion of a city of considerable size.

This paper led to much interesting enquiry, and light was thrown on the subject by comparison with blind men. Several members referred to cases of the latter whose history was known to them. The President thought that there was no doubt that the results, both in men and the lower animals, were dependent not only on greater acuteness of the other senses, but on the greater amount of attention paid by the mind to the data furnished by the former. It was to be remembered that improvement in the senses, whether in the blind or others, was largely to be referred to the brain

---

* "Comparative Psychology: Its Objects and Problems," *Popular Science Monthly*, March 1887.
† *Science*, vol. ix. Nos. 217, 222.

itself. This was especially clear on studying blind persons. The progress made even in walking under difficulties was owing, in the most successful cases, in great part, to superior brain development. The subject is of wide scope and of the very highest interest, but we cannot enter upon a further discussion of it now. At our January meeting Mr Pease reported some observations on a black-and-tan bitch. He had proved to his entire satisfaction that this animal understood the meaning of certain words perfectly well, in so far as could be judged by her actions. Thus she never confounded such words as " breakfast," " dinner," " supper.' It will be seen that we have given the question of the extent to which the dog understands words a good deal of examination. It merits the closest study, for unquestionably the magnitude of the gap between man and the lower animals is owing to the capacity of man to use, and his actual employment of, language. But I must repeat what I said in my last year's address, that it is more than likely that we much under-estimate the capacity of animals to communicate with each other by a language of their own.

Unlike the dog, the cat has received very little attention and consideration from man. There are many reasons for this neglect, but not least in significance is the fact that puss is no flatterer; the dog adapts himself to every caprice and whim of his master, but the cat is always herself. To understand her thoroughly, to see her at her best, she must be manipulated as a delicate piece of mechanism, and treated in the very kindest fashion. When so dealt with, the cat proves to be by no means only a comparatively untamed embodiment of certain strong instincts. I have maintained, and supported the opinion by some evidence, that the intelligence and possible good qualities of the cat have been much under-estimated.

The same may be said of another animal that has been not only neglected by those interested in the study of animal intelligence, but misrepresented in general. I mean the pig. What would the dog be to-day if he had, for hundreds of years, been valued only for his flesh, and kept exclusively to be fattened for food? The hog is charged with being dirty, stupid, and obstinate. Why should an animal, overburdened with flesh and fat, and consequently a sufferer from the heat of summer, be so much blamed for betaking himself to a pool even if muddy? Man is largely responsible for enforcing conditions involving filth on the hog. That this animal is not lacking in intelligence has been shown by his having been taught to hunt like a dog,* and by an interesting case reported to our Association by Mr Frank Miller. The animal was of the Chester White breed, had been trained at the age of four months, knew his name, would dance to music, go seek when told, lie down, and obey other commands. You will at least agree that the hog is worthy of further consideration at our hands. Circumstances were mentioned by one of our members which pointed very strongly to the possibility of hogs having hibernating capacity. This subject is of great physiological interest, and not without its bearings on Comparative Psychology; any new light on the subject, so far as animals, especially, are concerned, would be very welcome.

We have had a few communications on the intelligence of our domestic grazing animals. Mr Torrance had observed that sheep had acted, on different occasions, as if they were aware of the approach of storms still distant. Their behaviour, in seeking shelter, had been coincident with changes in the barometric pressure.

---

* "Animal Intelligence," by G. J. Romanes, p. 339.

Numerous reports from the sites of recent earthquake shocks by observers of unquestionable reliability have shown that many different kinds of animals were sensible of something abnormal, which caused in them manifestations of uneasiness or fear, some seconds before anything unusual was noticed by man. As I hope to show on some other occasion, such indications of acute sensibility, on close observation, throw much light on certain vexed questions in the science of Comparative Psychology.

Among wild animals we have had several short but interesting communications on the gopher of the prairies; also a very carefully prepared paper from Mr Harris, giving the results of his own investigations of the beaver and his work in the Canadian North-West.* Confirmations of these observations and additions thereto were offered by another member who had much experience of life on the prairies. It becomes very clear that the beaver is not only an animal of strongly pronounced instincts, but of great capacity to adapt itself to circumstances (plasticity of instinct, etc.). I again raise the question: What is the mental difference between the performances of the beaver and those of a man with marked genius for engineering operations, apart from all training? Only prejudice can prevent us seeing that this is a case of the highest suggestiveness, and it is, to me, replete with instruction. The time would fail me to attempt to even indicate to you how far-reaching is such an enquiry. I cannot help thinking that man would both understand himself better and have a truer insight into the inner life of the so-called inferior animals if he could get rid of some of his conceit, and regard himself rather as of, than

---

* This paper has since been published, by request of the editor, in the *Canadian Journal of Fabrics*.

apart, from the rest of the animal creation. The achievements of the nineteenth century are great; so also is its conceit.

The study of the apes and monkeys, on account of undoubted physical and mental resemblances to man, is naturally of the greatest interest. Accordingly, a communication from Mr Clement on a monkey he had kept under observation was welcomed by the Association. This creature's curiosity, observing powers, retentiveness of memory, and confidence in his owner, in contrast with a shyness towards strangers, were pronounced. His power of imitation, it was thought, had much to do with his mental progress. The superiority of this monkey, as in other cases, was evidenced by his capacity for education. As Mr Clement well observed, there was much in the creature's behaviour that suggested the child. The President had, in the case of this individual, verified Darwin's statement that monkeys have an instinctive fear of snakes. When this animal was offered a dead snake in a paper bag, he cautiously peeped in and then ran away in terror, nor could he be induced to go near the bag again. I may mention, incidentally, that there is now, in Central Park Menagerie, New York, a remarkable chimpanzee, of an intelligent expression of countenance so human-like as to be positively startling. If now he could but stumble on speech, what then? It seems not unlikely that the superiority of the monkey's brain over that of other animals may be owing in part to the use of the fore-limb as a hand. It has even been suggested that the greater brain-weight of man, compared with that of woman, may, in part, be the result of his more pronounced muscular development.

We have endeavoured to throw some light upon the question as to whether any animals have a special aptitude in finding water. There is a certain amount of evidence in the affirmative as regards frogs, turtles, and allied animals. A member, well acquainted with life on the plains, referred to the fact that thirsty travellers are accustomed to follow a "buffalo rut" in the confident hope of finding water. We need more exact information on such subjects very much.

Turning now to that most faithful servant of man, the horse, we must confess to having made less progress than in the study of the dog; and I would suggest that the reason is partly to be found in the very fact, that this animal is a servant, rather than a companion, of man. The whole nature of the horse is restrained and modified so that he may be adapted to human uses, and as a result, we fail to see him in his true nature, as a freely developing animal. The horse has become, to a large extent, a living automaton; as such, he is an interesting evidence of the dominance of one intelligence and will over another, but the real nature of the animal is, in consequence, much obscured. We have had, however, many interesting communications on the horse. At an early meeting of the Association Mr Dawes presented the formulated opinions of an expert trainer. Among the most important of these are the following: Horses know their names; they recognise each other after long separation; the development of high-speed trotting is largely dependent on the intelligence displayed in the training; horses exhibit judgment in the choice of track and in jumping hurdles; some horses neigh when the groom is seen at the feed-box or the water-tap; some even attempt to turn the tap; horses frequently endeavour to throw vexatious or unskilful riders. I shall be glad to communicate to the Association the results of an important interview I had during the past summer with an eminently successful trainer. His experience confirms and amplifies views I have on more than one occasion expressed before you.

Several members stated that they had noticed that horses have a special dread of bears, and could scent them at a great distance. Some horses were also afraid of a fur coat. No one, however, seemed to be able to explain why this fear should attach to bears more than to other ferocious animals; one would expect, in fact, that there would be much more danger to the horse from wolves than bears. Is this the remnant of a once powerful instinctive fear? At a later meeting Mr Dawes continued the subject of equine psychology. He explained that a mare in his possession had learned to overcome, in succession, different mechanical contrivances, such as buttons, fastenings, etc., which had been placed on the feed-box to prevent its being opened. Mr Dawes also instanced the behaviour of one of his horses, which showed considerable intelligent association of ideas. This animal was accustomed to being driven to the railroad station, and on certain occasions, on hearing the whistle of the approaching train, had started off on his own account, and after the train had left he had returned home. The ability of horses to remember incidents, sometimes trivial, after the lapse of years, was testified to by several members.

Mr Ferron has also favoured us with some interesting facts in regard to the intelligence of trotting horses; and the Principal of the College has pointed out instances of equine sagacity of a very striking kind. Horses had even come to the College hospital for treatment of their own accord. While all our domestic animals amply repay good treatment, of none is this more true than of the horse. To take advantage of this animal's gentle, sensitive, plastic nature to subject it to abuse, is the part only of a savage, and not of a civilised human being. It is not to be forgotten that ill-treatment of the brutes reacts on the moral nature of the man that is guilty of it; in injuring them he injures himself far more.

But the time fails me. This is necessarily a very inadequate account of our work up to the present—a mere sketch—but I hope it may suffice to encourage old members and to arouse the interest and enthusiasm of those now entering to fill the places of the men who have left us, and whose efforts in this cause we must gratefully remember. There is one thing which cannot in any way be represented to others, and that is the delight we have experienced in meeting together to discuss the inner and, unfortunately for us, so much hidden life of those beings that we have learned to regard with more and more respect, and to consider fellow-creatures. It will take some time to educate the public mind up to the point of realising how much these animals are really deserving of serious, respectful consideration. To the enlightened veterinary surgeon must we especially look for an improvement of the condition of our domestic animals, and in no way can this be accomplished more effectually than by learning their true nature and making that known. We wish to reach only the truth. No cause is in the end advanced by over-statement of the facts.

Sensible rather of how much is still to be done, than satisfied with our past progress, we renew our enquiries in the firm belief that an honest, humble search after truth will never be in vain.

PSYCHOLOGY AND COMPARATIVE PSYCHOLOGY.*

It is now more than ten years since I suggested to a few of the students of this Faculty of Comparative Medicine that it might be interesting and profitable to band together for the study of the psychic nature of animals, particularly those animals with which we are brought into daily contact.

In December 1885, at a meeting called to consider the subject, it was unanimously decided that a Society should be formed to study Animal Intelligence as best it could. Practically all the students, and those teachers more immediately connected with the work of this Faculty, joined the Association and entered into the new project with enthusiasm. It was early decided that only material obtained either at first hand, or from the most reliable sources, should be brought before the Association, and that principle, the wisdom of which will not be questioned, has been acted upon throughout.

Whatever the value of the papers and discussions which have engaged our attention, it may be fairly claimed that the facts upon which they have been based were beyond question. The first essential in any student of nature is a strong desire to know the truth, and, therefore, a great respect for exact observation at the outset. While theories change—and this is inevitable owing to the imperfection of our grasp of many-sided truth—a fact is always a fact. The patient collection of facts, so well illustrated by the illustrious Darwin, when

---

* An Address delivered to the Association for the Study of Comparative Psychology in Montreal, 1896.

theorising without very great regard to them was so tempting in framing explanations of organic nature, is a work that the world long undervalued, and the importance of which it is to be feared all psychologists at the present day do not adequately appreciate.

In this, at all events, our unpretentious Association may claim to have trodden in the safe path. At the end of our first decade of existence it may be profitable to review what has been accomplished. It could scarcely be expected that the members of this Association, being for the most part undergraduates, whose time is largely taken up with professional studies, should be able to make elaborate original researches worthy of publication. From the first, however, our proceedings have been given to the public in condensed form by the local Press, and evidence has been abundant on every hand that one of the results has been an altered attitude of mind on the part of many intelligent persons in this city towards the animal world about us, notably our domestic species. This is not a work to be despised, for the welfare of our fellow-creatures lower in the scale is largely dependent on the views we entertain of their psychic nature.

It is surely not to be supposed that such studies as have engaged the members of our Association are without a value of a professional kind; for in the handling of sick animals, in diagnosing their exact condition, in appreciating their sensations, and generally in understanding their entire nature, the man who observes and reflects on such things must be more competent as a veterinarian, other things being equal, and certainly a more agreeable visitor to both patients and clients.

But it is difficult, in my opinion, to over-estimate the good to the individual who, in the right spirit, studies animals. A frame of mind is established which, even when one exaggerates animal intelligence, is rarely practically harmful—often the reverse—and nearly always begets sympathy and modesty.

Psychology has passed through great changes during even the last decade. Now almost every college in America of much importance has its Chair of Psychology, and many colleges are provided with psychophysical laboratories. In America alone there are two periodicals devoted to this subject, and at last pedagogical institutions are attempting to found the training for teachers on the laws of the mind, *i.e.* on psychology. In fact no recent educational movement has been more widespread in its influence, or more rapid in its development, than the modern psychology.

The scope and methods of the science have also changed. While none the less introspective, it has become more objective. The allied science of physiology owes something to psychologists, notably in the direction of a more complete and accurate study of the senses, and keen criticism of positions assumed by physiologists in regard to the central nervous system.

The psychologists have borrowed freely from the realm of mental and nervous disease, all of which marks a new departure from which not only psychology, but physiology and practical and scientific medicine, must benefit.

It is usually a hopeful sign when methods of exact estimation begin to be applied to any science. There has been much diversity of opinion as to the extent to which this can be or has been successfully done in psychology. In the opinion of one of the most accomplished workers in this department of the science, who occupied the Presidental Chair at the last meeting of the American Psychological Association, there can be no doubt about the value of such methods and their application. He says: "I venture to maintain that the introduction of experiment and measurement into psychology has added, directly and indirectly, new subject-matter and methods, has set a higher standard of accuracy and objectivity, has made some part of the subject an applied science with useful applications, and has enlarged the field and improved the methods of teaching psychology."

But what shall we say of the *status* and prospects of Comparative Psychology? The works of Romanes were well known prior to the beginning of the last decade. They may be considered as marking about the first serious attempts to treat the subject of Comparative Psychology in a truly scientific spirit, and in a form accessible to the intelligent portion of the general public.

Much later appeared the books of Professor Lloyd Morgan—works which possess the charm of unusual clearness. If Romanes was open to the charge of claiming too much for animals, Morgan is certainly cautious enough to please the most conservative, unless it be those who deny true intelligence to animals entirely.

It is a hopeful sign of the times in psychology that a professor of philosophy, Dr Carl Groos, of Giessen, has found material for a book of considerable size on the play of animals, a subject which has been treated by him with interest, learning, and critical acumen.

Animal intelligence is more and more attracting the attention of the professed psychologist and biologist, and that both realise the difficulties of the subject, while its importance is acknowledged, is of good omen. Comparative psychology is now beyond the stage of neglect and contempt, though there are those who seem to think that before we can judge of the mental processes of animals, much greater progress must first be made in the study of the human mind; in other words, they would take their standards, their criteria, from human psychology. That we must in the end find the clue to interpretation from ourselves there is no doubt, but is it not the fact that every complicated subject has been advanced by studies on a lower plane and by the process of comparison? Anatomy and mammalian embryology would scarcely be worthy of the name of sciences to-day but for studies conducted on simpler forms. Do not psychologists sometimes forget, as anatomists long did, that the human is scarcely to be comprehended apart from the study of simpler creatures? Should we not look at psychology as the naturalist now does at zoology, and endeavour to discover the various grades in psychic processes, if such there be, and it is only, so far as I can see, by comparative investigation that their existence or non-existence can be established.

To do such work at its best requires a knowledge of both biology and psychology, and an intimate acquaintance with the ways of animals. Closet lucubrations cannot be expected of themselves to advance comparative psychology very much.

Might not human psychology be made more objective still, and is not the amount of wheat garnered much out of proportion to the quantity of sheaves brought to the thresher? Has individual psychology received the attention it deserves? Might not the inductive method be more fully applied to psychology? I have long been convinced that differences for races and for individuals have been insufficiently recognised in physiology, and at last there seems to be a reaction against the former reckless leaps from frog or rabbit to man.

The physiologist cannot, however, afford to ignore the frog or the rabbit even when his goal is man; nor, if I may venture to express an opinion, can the psychologist do so either without some loss—possibly great loss—to his subject.

I hope to see published, in the next few years, detailed studies on many individual human beings of both sexes, and also on individual animals. We must have more facts for our conclusions. The departures of French psychologists are very welcome, whatever the final outcome may be. It cannot be doubted that the study of hypnotism, double personality, and morbid states of various kinds has greatly advanced our knowledge of the normal man and his fellows lower in the scale; and I should be disposed to say that the investigation of the psychic processes of animals aids in the comprehension of even such abnormal states as those to which reference has been made.

At the recent great Psychological Congress at Munich there was, among others, a Department for Comparative Psychology; and an Endowed Lectureship on this subject has recently been established at Aberdeen, so that it is clear that in this, as in other directions, the world is moving.

If my view is correct that we are in need of vastly more facts and observations, then is there room for many workers. The experimental has a wide range of application in Comparative Psychology, and as yet but little has been done. In this direction, as I have urged for years on our members, we could do much to advance the subject we have at heart.

It has been my happy privilege to attend every meeting of this Association held since its foundation, and, reviewing the work of the past ten years, I feel that, although it has been a humble one, the Society for the Study of Comparative Psychology in Montreal has not existed in vain.

PART II.

SQUIRRELS: THEIR HABITS AND INTELLIGENCE, WITH SPECIAL REFERENCE TO FEIGNING.

WITH AN APPENDIX

UPON THE CHICKAREE, OR RED SQUIRREL.

I.

UNTIL recently the habits of animals seem to have been considered simply as interesting manifestations of their life, but without any special reference to their relations to the intellectual part of the creatures concerned. But unless we assume that animals are devoid of mind and true intelligence—an extreme and untenable position—there must be a possible science of Comparative Psychology, as there is of Comparative Anatomy and Physiology. The study of animal intelligence is possible, interesting, and important, whether we regard man as derived from some lower form, and his intellectual as well as his physical being the result of evolution, or whether we consider that man stands wholly apart in origin either as to body or mind. In the latter case the study of the lower forms of mind affords a useful contrast with its highest development as seen in man; in the former we aim at the construction of a ladder by which we may climb from the simplest manifestations of consciousness to the highest performances of the most gigantic human intellect.

I have selected the study of squirrel psychology as

the subject of this paper, because so little seems to have been written on the subject; because these animals are open to the observation of every one; and chiefly because I have been able to give special attention to them myself. Their habits will be considered principally, but not exclusively, from the psychological standpoint, and I shall apply the comparative method, making such references to the habits and intelligence of other rodents as seem to throw light on those of the squirrel. While some attention has been paid to other species of squirrels, my studies have been chiefly on the Ground Squirrel (*Tamias Lysteri*) and the Red Squirrel (*Sciurus Hudsonius*).

These species, in many respects, form a contrast to each other. The Chipmunk, Chipping Squirrel, or Hackee, has his abode underground in a specially constructed burrow; the Red Squirrel, or Chickaree, lives in nests in trees, and the intelligence of the latter seems to be altogether of a much higher order than in the Ground Squirrel. This was abundantly illustrated in my experiments with an ordinary wire rat-trap having a spring door. The trap was scarcely laid down near the haunts of the Chipmunk before one entered it, in fact before my eyes, and there was never any difficulty in securing as many as were wanted. On several occasions, when one had escaped in the room, on placing a small apple in the cage, the creature re-entered it almost at once.

Very different was it with the Red Squirrels; at first they entered the trap, but not afterwards. They approached it, sometimes two or three together, ran round it on the upper rail of the fence on which it was placed, or sat on the top of it—in short, did everything but enter it; all the while seeming to enjoy the whole greatly.

Having secured a couple of Ground Squirrels in the

manner described, I kept them under observation for the period during which they survived, viz., one for about a month and the other for between two and three months. From the first one of them seemed to take more kindly to his new surroundings than the other; one appeared shy and dull, while his fellow seemed as happy as any Chipmunk might be. They were captured in September, and it has often occurred to me that their habit of hibernation had something to do with the behaviour of the one, though we should expect that, in such a matter, both would be equally or considerably affected. The degree to which, while retaining their original habits, the latter became modified in confinement, furnished me with an interesting study, and suggested many problems. My experience does not agree wholly with that of Audubon and Bachmann, who say in their "Quadrupeds of North America": "We are doubtful whether this species can at any time be perfectly tamed." The one of my Chipmunks that survived longest became, in a short time, so tame that he would eat from the hand, and even looked to be fed in this way. True, any noise, or any unusual movement, might startle the creature, when he would make the quick dart away so characteristic of the species in the wild state, but from this he very quickly recovered, and the tendency to be thus frightened grew less and less. The authors referred to also state that " they appeared to have some aversion to playing on a wheel, which is so favourite an amusement of the true squirrels."

This does not at all agree with my observations, for though at first my Chipmunk was apt to be startled when he found the revolver of his cage moving on his entering it, he soon got used to it, and delighted in it as much as any squirrel could—in fact, he used it by night and by day, manifesting an ability to control it which speaks much for the readiness with which such animals adapt themselves to new and difficult movements, and which shows how highly developed those parts of the brain must be which are concerned in the balancing and kindred functions. I may here correct another statement of the same authors. They maintain that squirrels do not lap fluids as the dog and cat. From repeated observations I know this to be an error, so far as the Ground Squirrel is concerned at least.

It has usually been assumed that squirrels, and indeed most rodents, feed wholly on vegetable food, and that in those instances in which the contrary has been observed, there was evidence of a perverted or morbid appetite. Audubon and Bachmann, however, state that the Flying Squirrel (*Pteromys volucella*, Des.) has been caught in traps baited with meat. A number of writers,* especially within the past few years, have drawn attention to flesh-eating habits in several rodents, mostly under peculiar circumstances. Some interesting questions arise in this connection : (1) In how far is any rodent carnivorous when abundance of all the different kinds of vegetable food that the animal uses is at hand? (2) What is the relation between confinement and altered appetites ? (3) In how far are such altered appetites evidence of morbid or perverted conditions, and in how far simply the expression of physiological needs? The whole subject, I am inclined to think, might be placed on a broad and sound physiological foundation, but before that can be done, many accurate observations are required, and possibly also many series of experiments. If we may judge by the common house rat, rodents possess unusual plasticity as to feeding and other habits, and not less as regards their mental life. I found that my Chipmunk would

---

* *Science*, voL viii ; *Canadian Naturalist*, vol. iii.

take a great variety of foods, though the experiment of feeding with meat was not tried. He drank milk greedily.

There is one very peculiar habit, interesting from a physiological point of view, to be observed in squirrels in confinement. A writer in *Nature* (vol. x.) says:—

I have noticed that whenever it [the squirrel] cleans itself, after licking, it sneezes violently three or four times into its fore-paws, then rubs them, thus damped, over its fur." And this writer raises the question as to whether this habit, which he believes voluntary, was confined to squirrels. He does not mention what sort of a squirrel his own was, but I have noticed this behaviour as of the most frequent occurrence in my caged Chipmunk. It seems to me, on the whole, most natural to consider it a voluntary act of the same character, and possibly for a similar purpose, as clearing the throat in the human subject, or perhaps even blowing the nose; and I am the more inclined to believe that it is voluntary from the account given of the Flying Squirrel, as observed by Prof. G. H. Perkins, and recorded in the *American Naturalist* (vol. vii.). This writer states that on one occasion his squirrel lapped some ink, but shortly afterwards manifested disgust, and indulged in violent sneezings. Under these circumstances it is difficult to understand, by anything in our own experience, how the act could have been reflex.

Speaking of the relative intelligence of squirrels, this writer says:—" I am inclined to believe that the Flying Squirrel does not possess as much intelligence as the Grey or Red, or some other species." From the entire account of the Flying Squirrel given by Prof. Perkins, I should suppose that the intelligence of this species and that of the Ground Squirrel are about on a par—the explanation of which will be considered later.

A question of much interest to the naturalist and psychologist, it seems to me, is the following, viz. to what extent the intelligence of animals that hibernate has been modified by this process, and in what directions? With regard to hibernation, so far as the squirrels are concerned, there seems to be great dearth of accurate observations—in fact, the same remark applies to the whole subject of hibernation, one of the most interesting in the whole realm of physiology. A number of observations are to be found scattered through the literature, but they are fatally lacking, in most cases, in precision of observation and accurate record of dates. From a short but valuable paper on the "American Chipmunk," in the *Popular Science Monthly* (vol. vii.), by Dr C. Abbott, we are led to believe that the Ground Squirrel spends some time in his burrow before hibernation begins, and that the food laid up is consumed in part before the winter torpor sets in, and more especially in spring before a fresh supply is obtainable in the usual way. Concerning the winter habits of other species, I have been able to learn nothing from any quarter that definitely settles the question as to whether they hibernate or not. Audubon and Bachmann (*loc. cit.*) state that as much as one bushel and a half of nuts has been found in a single hollow tree occupied by a Chickaree, or Red Squirrel. They also state that this species may have several hoards. From different remarks dropped by these writers, from what I have myself observed, and from the statements of Dr Bell in the valuable notes appended to this paper, I am inclined to the belief that the Red Squirrel, and some other species, do not regularly hibernate the whole winter through; but whether they hibernate at all, in the true sense of that term; whether they have short periods of hibernation followed by intervals of consciousness, during which they feed; whether they remain in a condition of partial torpor, with slowing of all the vital processes, and yet not in absolute insensibility and with cessation of respiration, etc.—all these questions seem to be as yet wholly undecided.

It has long been known that many cold-blooded animals hibernate and, under altered conditions, æstivate; it is further believed that among warm-blooded animals, besides bats, many rodents, and some allied animals hibernate. But when the matter is looked into carefully it is found that the term "hibernation" has been used in a loose and very plastic sense by different authors. It is highly desirable, therefore, that writers should state exactly to what extent the animal they describe as "torpid," "hibernating," or "in winter-sleep," deviates functionally from the normal; also, that the exact time of the observations be recorded. There is a certain amount of evidence that even birds, representing the highest type of activity, may possibly hibernate, and that many animals, not usually thus affected, may become so under exceptional circumstances — indeed that man himself, owing to peculiar states of the nervous system, may pass into a condition ("trance") having much in common with the hibernation of lower animals. I think it is very probable that, when the matter has been fully investigated, all degrees of cessation of functional activity will be found represented, from the normal daily sleep of man and other animals to the lowest degree of activity consistent with the actual maintenance of life. The Flying Squirrel is nocturnal in habits and exceedingly active, even in confinement, as Prof. Perkins (*loc. cit.*) has shown; but during the day-time it seems not to be correspondingly quick—in a condition, in fact, resembling somewhat that of a hibernating animal. The "diurnal hibernation" of the bat is not to be forgotten. I noticed that my Chipmunk invariably, after feeding, tucked his head down and assumed a more or less ball-like form, highly suggestive of a tendency to hibernation.

There are many questions that arise in connection with this subject, one of which bears directly on the subject of Comparative Psychology: How and to what extent is the intelligence of animals influenced by hibernation? It may be considered pretty clear that both the Ground Squirrel and the Flying Squirrel hibernate, and these are certainly among the lowest—perhaps are actually the lowest—in intelligence of the whole tribe. We know that struggle among higher animals develops mental adaptation and other forms of intelligence, and it is rational to suppose that those species of squirrels that do not hibernate throughout the winter, but endeavour to prevail over their surroundings, as well as to adapt themselves to them, should be more intelligent than those spending a large portion of each year in inactivity.

My Chipmunk, during its captivity, under certain circumstances, kept to his original habits, *e.g.* when a single nut was given him he would eat it immediately, but if several were presented at once he would hide them, one by one, in a corner of his cage, or, if sufficiently small, pack them away in his cheek-pouches. He did the same with cereal grains. When cotton-wool or web-like material was placed in the cage he manipulated it a good deal, but finally made a bed of it, in which he buried himself out of sight.

Within the last ten years attention has been called to "singing" in certain rodents, especially mice; but from numerous references in the literature it appears that "singing," or something analogous to it, has been noticed in a large number of rodents.* The well-known note of the Chipmunk, from which it has derived its name, is the only one I have heard from it. After studying a colony of Red Squirrels for some weeks last summer, I came to the conclusion that they have a capacity of vocal expression much greater than is commonly believed. Their usual "barking," or trilling, seems to be the commonest, the most instinctive, and not largely expressive of anything beyond general satisfaction; but, I found that, under excitement, there were many other tones, associated with great complexity of emotion, which I am not prepared to analyse, but which there can be little doubt the creatures themselves employ as a means of inter-communication. Under marked excitement, as the result of repeated interferences, I have heard a Red Squirrel so mingle tones of a musical kind that, a stranger arriving on the spot, would certainly have been deluded into the belief that he was listening to some bird, or rather to an excited pair of birds. The musical character of this combination, together with its continuity and complexity, would perhaps justify the designation "song." One of the writers on musical mice refers to their singing but little in certain instances, except when excited, which is a point of analogy with the Chickaree.

It would appear, therefore, that it is likely that throughout the order Rodentia a genuine musical appreciation and executive capacity exists, and, in some instances, in a very high degree; and that apart from this, there is also considerable ability displayed in the expression of states of emotion, at least, by vocal forms. Manifestly, the degree to which animals can express their psychic states—and especially in vocal forms—is a matter of the greatest importance, and I have already

---

* *See* especially *Nature*, vol. xv.; *Popular Science Monthly*, vol. i.; and the *American Naturalist*.

expressed my conviction that animals have a power of communicating with each other, altogether beyond what has been generally surmised. The subject is beset with great difficulties, and calls for the closest observations.

II.

I PROPOSE, in this second part of my paper, to discuss the subject of feigning in animals, and shall give, as a basis for my views in the case of the squirrels, an account of two Chickarees, in which such behaviour was strikingly manifested.

Case I.

I was standing near a tree in which a Red Squirrel had taken up a position, when a stone thrown into the tree was followed by the fall of the squirrel. I am unable to say whether the squirrel was himself struck, whether he was merely shaken off, or how to account exactly for the creature's falling to the ground. Running to the spot as quickly as possible, I found the animal lying apparently lifeless. On taking him up, I observed not the slightest sign of external injury. He twitched a little as I carried him away and placed him in a box lined with tin, and having small wooden slats over the top, through the intervals of which food might be conveyed. After lying a considerable time on his side, but breathing regularly, and quite free from any sort of spasms such as might follow injury to the nervous centres, it was noticed that his eyes were open, and that when they were touched winking followed. Determined to watch the progress of events, I noticed that in about an hour's time the animal was upon his feet, but that he kept exceedingly quiet. The next day he was very dull—ill, as I thought—and I was inclined to the belief, from the way he moved, that possibly one side was partially paralysed; but finding that he had eaten a good deal of what had been given him (oats), I began to be suspicious. Notwithstanding this apparent injury, that very day, when showing a friend the animal, on lifting aside one of the slats a little, he made such a rush for the opening that he all but escaped. On the third day after his capture, having left the sitting-room (usually occupied by two others besides myself) in which he was kept for a period of about two hours, I was told, on my return, by a maid-servant and a boy employed about the house, that some time previously the squirrel had escaped by the window, and, descending the wall of the house, which was "rough-cast," he had run off briskly along a neighbouring fence, and disappeared at the root of a tree. When asked if they saw any evidence of lameness, they laughed at the idea, after his recent performances before their eyes. For several days I observed a squirrel running about, apparently quite well, in the quarter in which my animal had escaped, and I feel satisfied that it was the squirrel that I had recently had in confinement, but, of course, of this I cannot be certain.

I believe, now, that this was a case of feigning, for if the injury had been so serious as the first symptoms would imply, or if there had been real paralysis, it could not have disappeared so suddenly. An animal, even partially paralysed, could scarcely have escaped as he did and show no signs of lameness. His apparent insensibility at first may have been due to catalepsy or slight stunning. But while there are elements of doubt in this first case, there are none such in that about to be described.

Case II.

A Chickaree was felled from a small tree by a gentle tap with a piece of lathing. He was so little injured that he would have escaped had I not been on the spot where he fell and seized him at once. He was placed forthwith in the box that the other animal had occupied. He manifested no signs whatever of traumatic injury. One looking in upon him might suppose that here was a case of a lively squirrel unwell, but events proved otherwise. He ate the food placed within the box, but only when no one was observant. He kept his head somewhat down, and seemed indifferent to everything. When a stick was placed near his mouth he savagely bit at it; but when a needle on the end of the same stick was substituted he evinced no such hostility. He made no effort to escape while we were in the room, but, on our going down to dinner, he must at once have commenced work, for, on returning to the room in half-an-hour, he was found free, having gnawed one of the slats sufficiently to allow him to squeeze through. With the assistance of a friend he was recaptured, but during the chase he showed fight when cornered, and finally, as he was being secured, I narrowly escaped being bitten. He was returned to his box, which was then covered with a board weighted with a large stone. Notwithstanding, he gnawed his way out through the upper corner of the box during our absence on one occasion shortly afterwards.

I think a more typical case of feigning than this one could scarcely be found.

The accounts of these two cases are based upon notes taken at the time, and this brings me to the most interesting, and at the same time the most difficult series of enquiries connected with the whole subject, viz. What, upon analysis, is this feigning in animals?

In how far is it instinctive, and in how far an intelligent and deliberate adaptation of means to an end under unusual circumstances? How did the instinct of feigning death and injury arise in the first instance? Has feigning been confounded with something else totally different, such as the results of fear, surprise, etc.? Is the expression, "feigning death," not misleading in itself? The matter is so intricate, and such diverse views have been entertained in regard to the subject of feigning, that it will be necessary, in order to arrive at a solution, to examine critically several of the views advanced.

Feigning death has been observed in many different genera of insects, in snakes, fishes, numerous birds, crustaceans, and several mammals.

In a most interesting account of experiments on certain animals, by Prof. Czermak, published in the *Popular Science Monthly* (vols. iii. and iv.), it was shown that in the crayfish, in hens, geese, ducks, turkeys, pigeons, the swan, etc., a state, which this writer recognised as having a physiological basis, but which he did not attempt himself to explain, occurs. In all these animals, under the influence of steady restraint of motion, or, combined with that, prolonged gazing at some object held just before the eyes, a condition of quietude and partial or complete unconsciousness was induced for a shorter or longer period, after which they regained their usual condition. In some of the animals the muscles became rigid, *i.e.* the cataleptic condition was induced.

About five years later Prof. Preyer gave the subject a thorough experimental examination. The starting point of all these experiments was the *experimentum mirabile* of Kircher, in 1646; Preyer seems to use the term "cataplexy" to cover what is now more commonly called "catalepsy," or "hypnotism." Preyer believed that the shamming death of certain species of the *Articulata*, when threatened by danger, was due to cataplexy. The condition was attended in certain animals by stupor, violent tremblings of the extremities, and other pronounced disturbances of function and psychic state. This writer then explained the condition, called by some "shamming death," by a sudden, powerful, unexpected, and unusual stimulus acting on the centripetal nerves, producing an emotion of fear which acts on the will, inhibiting it and producing stupor; "deathly terror," in a word, is the condition, and not feigning, according to Preyer.

The well-known physiologist, Heidenhain, performed many experiments, chiefly on the human subject, with a view of arriving at a physiological solution of these remarkable phenomena. He has framed the theory, that hypnotism is due to the inhibition of the cortical cells of the cerebrum, caused by the gentle prolonged stimulation of the nerves of the face, eyes, or ears.

Dr Clarke, in the *Popular Science Monthly* (vol. ix.), discusses the results of Czermak and others, and concludes that "they depend wholly and only on fear," for he maintains that the experiments succeed best in the wilder individuals of the species. But Dr Clarke is scarcely consistent, for he points out in the same paper that animals cease to struggle because they find it useless, and this he ascribes to intelligence.

Dr D. W. Prentiss, in the *American Naturalist* (vol. xvi.), examines the matter from the physician's point of view. After referring to the "dancing," "convulsive," and "laughing" manias, and to certain phenomena in animals like those already described, he concludes that the factors entering into the phenomena of Czermak and others are fear, dissembling, curiosity, training, changes in the condition of the blood (deficiency of oxygen from restrained chest movements), and imitation.

To my own mind all these explanations are partial and inadequate. That terror, surprise, etc. are in no sense essential for the induction of hypnotism is sufficiently evident from Czermak's experiments on pigeons, which could not be put into this condition by mere restraint, but only after uniting with this steady gazing at a near object. Again, it is well known that the human subject can be hypnotised by the latter means alone, as Heidenhain first attempted to show. The latter's explanation, though perhaps as good as can be given in the existing state of physiological knowledge, does not apply evidently in its present form to animals in which the cerebrum is not developed, as in insects and other invertebrates. The view of Dr Prentiss has the merit of breadth, but manifestly some of his factors, as training, imitation, etc., cannot apply to the hypnotic condition when first experienced, at least in the lower animals.

Notwithstanding the inconsistency in Dr Clarke's article, he is probably quite correct in explaining the quiet of animals, when restrained, in many cases by an intelligent perception that struggle is useless. I have, myself, frequently noticed, when controlling rabbits in the laboratory for the purposes of observation, that so long as there was no part of the fastenings loose, they remained quiet without any attempt at freeing themselves; but, if only a single limb became the least free, then a general struggle began. But such an explanation will not suffice when a greater or less degree of unconsciousness supervenes.

It may, I think, be said that the phenomena included under such terms as hypnotism, cataplexy, etc. are due to influences reaching the nervous centres, unusual either in quality or intensity, or with an altered relation as to frequency of repetition when compared with those associated with the ordinary experiences of the animal.

When we fully understand the physiology of sleep, we may then be able to give a final and satisfactory explanation of these phenomena, but scarcely before. However, I venture to assert, that most, if not all of the phenomena of hypnotism, may find psychological realisation in the experiences of every individual human being, if he will but observe himself closely enough over a sufficiently long period of time.

Turning now to feigning death. This subject did not escape that great master of close observation, Charles Darwin. He says, in his " Essay on Instinct " (now published as an Appendix to Dr Romanes' work, " Mental Evolution in Animals ") :

" Insects are most notorious in this respect. We have amongst them a most perfect series, even with the same genus (as I have observed in *Circulio* and *Chrysomela*), from species, which feign only for a second, and sometimes imperfectly still moving their antennæ (as with some *Histers*), and which will not feign a second time however much irritated, to other species, which, according to De Geer, may be cruelly roasted at a slow fire without the slightest movement ; to others again which will long remain motionless, as much as twenty-three minutes, as I find with *Chrysomela spartii*."

Darwin speaks of such feigning as instinctive. Romanes (*loc. cit.*) believes it instinctive, but thinks cataplexy may have been of much assistance in originating and developing it. Both of these writers agree, however, that instinct has been perfected by natural selection.

If this shamming death, or rather assuming the position of the dead, were really of benefit to the animals, such an explanation might be valid if natural selection be admitted at all. On the other hand, Darwin has shown that the position assumed by the shamming insects "in no instance was exactly the same " as that of the dead insects, and in many cases it was as unlike as could be. The question then arises in my mind: May not this condition assumed by insects be a peculiarity with which natural selection has nothing to do—a sort of imperfection of their nervous system, if it exposes them to enemies, the reverse if it conceals them—at all events, not necessarily connected with natural selection, for animals survive in spite of peculiarities and imperfections? In fact, the conception that any animal is perfectly adapted to its surroundings is unwarrantable, otherwise such an animal should continue to live *in perpetuum*.

Preyer would ascribe the so-called shamming death of insects wholly to cataplexy, which seems highly probable. Couch, who is quoted by Romanes, would explain certain behaviour of wolves, foxes, and some other animals, usually set down to deliberate feigning, by an effect analogous to cataplexy. He thinks their senses are stupefied by surprise, terror, etc., so that they are unable to escape.

The transfixing effect of fear in man has been well described by the poets, including Shakespeare himself:—

> "Whilst they, distill'd
> Almost to jelly with the act of fear,
> Stand dumb, and speak not to him."

Romanes inclines to give weight to the views of Preyer and Couch so far as vertebrates are concerned. He says: "A fox would never have so good a chance of escape from an enemy by remaining motionless as it would by the use of its legs." But if man is to be reckoned among the enemies of this animal, then, according to instances given by Romanes in the same chapter, foxes have escaped from their enemies by feigning death.

I have often noticed how one dog has escaped the attack of another by lying down and assuming an attitude of complete surrender (*see* "Youatt on the Dog," Amer. Edit., p. 34.) Even dogs would not be inclined to worry a fox apparently dead. And what of the feigning of the opossum ? Romanes finds a special difficulty in this subject, because, as he says: "On the one hand, it is obvious that the idea of death and its conscious simulation would involve abstraction of a higher order than we could readily ascribe to any animal, and, on the other hand, it is not easy otherwise to explain the facts."

I cannot help thinking that this difficulty is a sample of those we make for ourselves by attempting to define and classify where Nature has left things complex and unsusceptible, of the sort of simple analysis after which Romanes and others are, in this instance, striving.

If there is a vertebrate animal in which the feigning of death is an instinct, as pure as such an instinct can be, that animal is the opossum (*Didelphys virginiana*, Shaw). If invariability of behaviour under similar circumstances be essential to an instinct, then the opossum's feigning is instinctive. From the account of a writer in the *American Naturalist* (vol. vi.) we learn that a Turkey Buzzard (*Cathartes aura*) may run upon an opossum and, after flapping his wings a few times over him, the opossum will go into a "spasm," and the buzzard proceed to pick out its eyes, and "generally take a pretty good bite from its neck and shoulders." From all that I have been able to learn of the behaviour of this animal in the presence of such circumstances as lead to its so-called feigning, I have been led to conclude that it is really largely, if not wholly, a condition allied to, if not identical with, Preyer's cataplexy; but no one seems to have given the subject that accurate examination necessary for a solution, in this, perhaps, the very best animal in which to test it. The creature is abundant, and could be captured at any age and degree of development. In this case, as the animal is poor in resources of escape by flight, etc., the instinct may be valuable to it, but from the above account evidently not always.

The general intelligence of the animal is low, for it will readily enter traps laid for it. I am the more confirmed in the above-stated view of the case,* and indeed of the extensive prevalence of such nervous phenomena, from an examination of an account of the behaviour of a Turkey Buzzard, given by Dr Prentiss in the paper to which reference has already been made. This writer states that, having winged a buzzard, on coming up with it the creature lay on its side as if dead. Believing it really was dead, he thrust it into his game-bag, brought it home, and threw it down in his yard, limp and apparently lifeless. A little later it was found running around, but on being approached it acted as before, and with each shamming it "disgorged," to use the writer's expression. He further states that after a while it would only disgorge and hiss. Now, on comparing this "disgorging" with the phenomena described by Preyer, as witnessed in his animals that were truly hypnotic, I feel quite persuaded that this case of the buzzard is explicable by the facts of hypnotism, especially as the symptoms disappeared largely on familiarity with the surroundings: it was not a genuine case of feigning. The author of the account does not himself clearly indicate his view of the case.

But Romanes, while inclined to the theories of Couch and Preyer as a partial explanation, adduces from the

---

* Since writing the above I have been pleased to find that Dr Charles C. Abbott has given the so-called feigning of the opossum a careful, one might say, experimental examination. He has discussed the subject in his work, "A Naturalist's Rambles About Home," and has been led to form conclusions similar to my own.

writings of others instances of feigning in monkeys which place it beyond doubt that animals may consciously and deliberately feign; yet he regards the matter as one of great difficulty. Unquestionably it is, but I must again express my conviction that Romanes has imported into the subject difficulties which are not in the nature of the case present. First of all, is it at all essential to "feigning" either death or injury that an animal should have, as Romanes supposes, the abstract idea of death at all? It is to be remembered that in these cases the animal simply remains as quiet and as passive as possible, which is in accord with all an animal's experiences as to escape from danger by any form of concealment. We have all degrees of this. The little Chipmunk, when a hawk is at hand, squats, if on a fence; if near its burrow, rushes in, according to Dr Abbott (*loc. cit.*). It is within the observation of all, that a cat watching near a rat-hole feigns quiet; in like manner a dog, desirous of capturing the fly that has been tormenting him, feigns apparent unobservance or unusual inactivity. I suspect that a human being, suddenly finding himself in danger, may, and often does, exercise a similar control without any abstract notion of death. Indeed, the extent to which the abstract in this sense enters into the psychic life of men, if we except the higher class of intellects and persons well educated, is much less than writers have been wont to believe. A great part of the whole difficulty, it seems to me, has arisen from the use of the expression "feigning death." What is assumed is inactivity and passivity, more or less complete. This, of course, bears a certain degree of resemblance to death itself.

Returning, then, to the case of my feigning Red Squirrels, I should be inclined to explain their behaviour somewhat as follows:—

By inherited instinct, as well as by all those life experiences which had taught them that quiet and concealment of their usual activities were associated with escape from threatened evils, these little animals were naturally led, under the unwonted circumstances of their confinement, to disguise, in an extraordinary degree, their real condition, and even to imitate an unusual and unreal one. The mental process is a complex of instinct pure and simple, with higher intellectual factors added, and the cases of these squirrels, thus feigning, are among the clearest that, so far as I am aware, have ever been recorded. The adaptations to effect escape prove that there was the employment of intellectual processes of a pretty high order, possibly too complex, however, for analysis with safety, but not beyond realisation in our own consciousness, and without the employment of any abstract idea of death.

That, however, the hypnotic element may play a part in the apparent feigning of death by squirrels seems clear from a case communicated to me by a student of the Montreal Veterinary College, Mr Craig. He had caught a Chipmunk and placed it in a box, to find in a few moments that it was lying as if dead. Giving the creature liberty to escape, it presently did so. On recapture the same followed. Considering the relatively low intelligence of this species of squirrel, and taking into account the case that Dr Romanes mentions of his watching an apparently feigning squirrel he had caught when he found that it had really died of fright, it seems to me, upon the whole, most reasonable to attribute the behaviour of the Chipmunk in question to cataleptic or allied effects.

It thus becomes manifest how varied, and also how complex, these cases of so-called feigning may be. The subject is all the more interesting because it shows that there is much that is common in the psychic life of human beings and that of the lower animals. It places the study of their habits and intelligence on a higher plane, and furnishes new motives for extending our enquiries and attempting to give unity to our conception of nature in this as in other domains.

Most remarkable evidence of high intellectual capacity has been furnished by the conduct of elephants under surgical operations, as instanced by Romanes in his "Animal Intelligence;" and Principal M'Eachran has assured me that both dogs and horses have shown a similar intelligence by coming, of their own accord, to his veterinary hospital to have injuries treated, after having been there and experienced the benefit therefrom. Dr G. P. Girdwood, a few days ago, gave me an account of what appeared to be a similar manifestation in a Chickaree but recently caught, though in this case so much, perhaps, cannot be claimed. This Chickaree submitted, soon after being caged, to having parasites removed from the skin, voluntarily remaining quiet during the act.

With regard to the psychological rank of the various species of squirrels, both from what I have been able to learn from the writings of others, and from my own observations, the Chickaree must be placed, I conclude, at or very near the top of the list. The Chipmunk and the Flying Squirrel seem to be, as already said, about equal in intelligence, and both much below the Red Squirrel, owing, perhaps, to the underground life of the one and the nocturnal habits of the other, possibly also to annual hibernation.

The wide geographical range of the Chickaree, as referred to by Dr Bell in the Appendix, of itself indicates great power to adapt itself to circumstances requiring intelligence, and it has been shown abundantly in this paper how the Red Squirrel can accommodate itself to new conditions and cope with emergencies.

To what, then, is the superior intelligence of this species due? In my opinion, partly to the fact that he has benefited by proximity to civilisation. While the Black Squirrel (*Sciurus niger*) seeks the depths of the forest, the Red Squirrel keeps near, by preference, to the abodes of man. Among rodents, none perhaps excels the domestic rat in general intelligence, a fact to be ascribed to this same human contact. Indeed, there is, perhaps, no group of animals that has long been near man that has not been more or less elevated in the scale of intelligence as a consequence, which, in turn, shows that the intellect of brutes cannot be wholly different from that of man. The applicability of this explanation to the squirrels is not so obvious as in the case of some other animals. The superior intelligence of the Red Squirrel is doubtless the resultant of a complex of factors which we can but imperfectly unravel, but from what I have observed as the result of actual experiment, I am forced to conclude that this creature can readily adapt itself so as to overcome the obstacles and avail of the advantages of man's civilisation; and I see no reason why, as a consequence of ages of inheritance of such naturally increasing capacity of adaptation or its results, the general intelligence of the species might not be raised. Such, however, probably constitutes but one element of a complete explanation.

APPENDIX

On the Chickaree, or Red Squirrel
(*Sciurus Hudsonius*, Pennant).

By Dr R. Bell, Geological Survey, Ottawa.

Geographical Distribution.—East of the Rocky Mountains the Chickaree ranges northward to near the verge of the forests, or to a line drawn from Fort Churchhill, on the west coast of Hudson Bay, to the mouth of Mackenzie River, and throughout the Labrador Peninsula, except the Barren Grounds which form its northern part. It is also common in Alaska. The rufous variety, universally known as the Red Squirrel, is abundant throughout the Canadian provinces and the northern, eastern, and middle states, extending farthest south along the Alleghany Mountains, or into Alabama. In the Rocky Mountains, and on the Pacific side, the varieties *Sciurus Douglassi* and *S. Fremonti* take the place of the more widely distributed form. An animal which can maintain a cheerful existence over such a continental area must necessarily be capable of adapting itself to a great variety of circumstances, as to climate, food, etc. The following notes will relate to the Chickaree in his more northern haunts:—

Food.—Northward of the zone of butter-nuts, beech-nuts, etc., the hazel extends a long way—say, to a line drawn from Lake St John (on the Saguenay) to Lake Athabasca, curving southward of James and Hudson Bays—and affords a large proportion of their food. Besides eating them constantly during the autumn, they store up considerable quantities for use later on.

But the seeds of the black and the white spruce constitute their grand staple in the north. By glancing at the map it will be seen that the extent of territory in which the spruces abound, to the total or partial exclusion of other food resources, is so great that it may be said that the area in which the Chickaree lives principally on the seeds of these trees forms more than half of the total range of the species, so that, taken as a whole, these seeds really constitute their leading article of food.

In old spruce forests in the north, the Chickaree is quite numerous, and almost every tempting log or hummock which commands a clear view all around (from which he can watch the approach of enemies while feeding) is covered with the scales of the numerous cones he has picked to pieces in order to get at the seeds. They evidently thrive on this diet, for their size and numbers, as well as their activity and audacity, are unabated.

HABITS.—Their mode of obtaining a supply of cones is ingenious. The cones grow principally at the tops of the spruce trees, and the largest and finest are always to be found there. The Chickaree selects a tree which, either from the steepness and density of its upper part, or from its leaning to one side, makes it certain that the cones, if detached, will fall to the ground; then he cuts off the heavily laden twigs and lets them drop. This is done with an impatient rapidity. Should a person be sitting quietly under a tree while one of these busy little creatures is at work at the top, he will see the bunches of cones come tumbling down in such quick succession that he might suppose half a dozen squirrels were at work instead of only one industrious little fellow. These bunches seldom lodge in the branches below, but should the squirrel, on his way down (after having cut off a satisfactory supply), notice one of them arrested in a hopeful position towards the extremity of a bough, he will sometimes run out and give it a second send-off. In climbing tall spruce trees for observations of the surrounding country, I have often noticed bunches of cones lodged where, if started off a second time, they would be certain to catch again in the thick branches before reaching the ground. The squirrels seem to understand the situation perfectly, and they leave such bunches to their fate, probably arguing that it would be easier for them to cut off fresh ones than to trouble themselves further about property lost beyond hope of profitable recovery—a piece of wisdom which the most successful business men have also learned to follow. The Chickaree, having thrown down a sufficient stock for a few days' use, proceeds to carry them, as required, to his favourite feeding-place, near by. I have occasionally noticed a squirrel feeding with a fresh cone lying beside the one he was actually dining off, as if it were waiting to be attacked the moment he got through with the first. They peel off the scales in succession, and nibble out the seeds with great rapidity. They leave their stock lying about under the tree, and only carry off one or two cones at a time. A little drying causes the scales to gape, and so facilitates the opening process.

In the northern regions referred to, in addition to spruce seeds, the Chickaree appears to feed a good deal on certain brownish, mushroom-shaped fungi. These they seem to prefer in a partially dried or decomposed condition, for they carry them up and leave them for a time on the flat, spreading branches on the sunny sides of spruce or other fir trees. I have sometimes seen one of them making off with a fungus nearly as large as his own body.

CONSTANCY OF CHARACTER.—No matter where you meet the Chickaree in the north country, he has precisely the same peculiarities of habit as elsewhere. In the depths of a dark spruce forest, which offers no temptations for a visit from even the few human inhabitants of these regions, and which it is pretty certain have never before been trodden by the foot of man, should you come suddenly upon a Chickaree, he greets you with the same saucy familiarity as he would if you disturbed him in a black walnut tree on the borders of Lake Erie. After scolding the intruder, with his head peeping round the trunk of a tree, should you throw a stick at him, or make a feint to run to the side he is on, he will merely dodge you to the other side and get up a little higher before reconnoitring you again.

WINTERING.—In the northern regions under consideration the Chickaree appears to pass the coldest part of the winter in nests in hollows under stumps, or in fallen trees, and the Indians say that they come out and run about on fine days in any month. They make nests—sometimes as large outside as a bushel measure—of moss, leaves, and a few small sticks, in the branches of trees in thickets, at moderate heights above the ground. These they appear to inhabit principally in the autumn and spring.

BREEDING.—The Indians have sometimes told me that the squirrels have their young in the nests just referred to, but I have not verified this statement myself. Their season of heat is said to be the early spring, just when the snow begins to melt. They rear but one family each year.

SIZE AND COLOURING.—Throughout the vast northern region of coniferous forests inhabited by the Chickaree, between the Atlantic Ocean and Mackenzie River, the animal presents great uniformity of size and colouring. It is considerably larger than the varieties to the south and west, and the colour, instead of being decidedly rufous above, as in the familiar Red Squirrel of civilised regions, is of a grey-fulvous tinge. Melanism and albinism, or any variation whatever, is extremely rare. I obtained a specimen at Athabasca Lake, which is pure white beneath, from the nose to the tail, the second half of which is also perfectly white.

ENEMIES.—The marten seems to be the principal enemy of the northern Chickaree, although they occasionally fall a prey to the lynx, and they also appear to keep a watchful eye on the mink, the fisher, and the weasel.

FEIGNING.—As to the Chickaree's habits of feigning, I may mention a case which came under my observation on Lake Superior. Being detained one day by a head-wind, my men amused themselves by capturing, alive, a mink and a Chickaree, both of which they put into an empty box with bars in front. The squirrel seemed to dislike the presence of the mink more than he did his captivity, and crouched in a corner with his head drooping and his eyes shut, as if very sick or totally disabled. After the mink had got over his first fright, and begun to take in the situation, he ventured to attack the squirrel, which immediately displayed great courage and activity, completely mastering his enemy for the time. Next morning, however, the poor squirrel was found dead with his throat cut.

HIBERNATION AND ALLIED STATES IN ANIMALS.

For a long time it has been known that many insects pass into a state of profound torpor during the winter season, from which they are apt to emerge, as seen in our house flies, when the temperature rises sufficiently high.

Snails are well known to provide against the approach of winter by closing up their shells, within which they sink into a protective sleep, and doubtless hibernation is a very widespread phenomenon among invertebrates.

There seems to be little doubt that in cold latitudes all reptilia and amphibia hibernate, and in warm countries æstivate. Nevertheless, definite investigations have been few. At the Philadelphia meeting of the American Association for the Advancement of Science of 1884, A. W. Butler made an interesting communication on this subject, giving some definite data, more especially in regard to the hibernation of the "box tortoise," "soft-shelled" and "hard-shelled" turtles, frogs, toads, newts, salamanders, and certain fishes, which may be found stated succinctly in the Proceedings of the Association for that year. His observations apply to Brookville, Ind., U.S.A. He concludes that: "In this part of the Ohio Valley, tortoises, turtles, toads, and frogs are regularly found hibernating; while, on the other hand, newts, salamanders, and many species of fish do not, as a rule, enter a torpid state."

There seems to be no doubt, however, that many species of fish do hibernate. Turning to warm-blooded animals (*homoiothermers*), it is thought that while the brown bear of Europe and the badger sleep most of the time in winter, they do not hibernate in the same sense as *e.g.* the dormouse. The black bear is believed to hibernate, though definite information about the winter life of this creature and other American bears generally would be very welcome. The hedgehog is regarded as a true hibernater the winter long. It is known that the tenrec of Madagascar sleeps for three months in his burrow during the hottest part of the year. It is, however, among the rodents that we must look for the most perfect hibernation; and the porcupine, the hamster, the dormouse, the marmot—and, as some think, the squirrels, are the best known examples. But if the hibernation of the bat is not the most perfect, it seems to have been about the best studied, for Marshall Hall's investigations of sixty years ago are still to be regarded as classic.

In consequence of reference to this subject in a paper on Squirrels, read before the Society in 1887, and my appeals for assistance in the study of this wide subject (in which one person can do comparatively little of himself, at least in ascertaining those modifications which apply for different latitudes and conditions), I have been able to obtain some definite information as regards the winter sleep of squirrels especially.

J. P. Bishop, Professor of Science in the State Normal School of Buffalo, N.Y., writes me:

"Regarding hibernation, I seriously doubt whether the Red Squirrel, in the latitude of Central N.Y., ever really hibernates. I have seen him out at all times of the winter, and in all kinds of weather, even when the thermometer recorded temperatures below zero and the air was full of snow. But he is much more lively in warm days, which he prefers for feeding. The Grey Squirrel is more sensitive to cold, but will come out upon almost any warm day in the winter."

I am also indebted to Mr W. Yates of Hatchly, Ontario, a capable and loving observer of Nature, for several notes. He says:

"Trappers are opposed to the belief that Red Squirrels hibernate, for they may be seen in the woods in the most inclement weather. They do not store up food in nests for themselves, but rob the dormouse or woodmouse, and kill him when they can."

This seems also to be Mr Yates' own view. He says that Flying Squirrels, when the hollow trees on which they take shelter are cut, at once betake themselves in the most lively manner to some other hollow tree in the neighbourhood, the location of all of which they seem to know beforehand. This scarcely argues a very deep sleep—if sleep at all.

As to the Chipmunk (*Tamias Lysteri*), which certainly stores up food in a burrow, he seems more in doubt, but is not inclined to believe that he hibernates for very long periods at all events. He has seen them out as late as 21st December. This observer has made some very interesting observations on a tame racoon (*Procyon Lotor*). This creature lived in a hollow log lined with straw, and "drowsed away the greater part of December and January, leaving any food placed before him unnoticed." The racoon is known to spend the greater part of the winter in hollow elm trees in this part of the country, and Mr Yates points out that the cutting-down of most of these trees resulted in the racoons betaking themselves to underground burrows, including those once occupied by foxes.

He says of the Woodchuck (*Arctomys Monax*):

"Our Marmot hibernates sooner than the bear, racoon or Chipmunk. Towards the last of October he deserts his burrow in the fields for one in the woods, choosing a dry, sheltered ridge, and is never seen out till mild weather returns."

On this point Prof. Bishop writes me:

"Popular belief puts the time of hibernation of the Woodchuck from 1st October to May. I have seen them out well into November, and have known of their being caught in traps on 1st January. I also saw the tracks of one myself on 2nd January. From that time until May I have run across them or their tracks after a warm day."

Mr Yates expresses a positive opinion. He considers that their winter sleep is of the most profound character, for he says: "The state of unconsciousness is death-like." As to bears, he simply reiterates the belief that they hibernate only when food is not obtainable, in which view, in the light of my own investigations, I see nothing impossible.

My studies on the torpor of cold-blooded animals (*poikilothermers*) have been little more than casual observations; on the bat they have been more extensive, still incomplete; but the condition of the Woodchuck, our American Marmot, during winter and summer in confinement, has had my close attention for over five years, almost continuously, by the help of other members of my family, especially of my wife, the time including the early morning and evening as well as the hours of the working day. Of the habits of this creature when in its natural surroundings I know but little from personal observation.

The specimen on which my observations and experiments were made during four successive years was kept in confinement for some time prior to its coming into my possession—how long I do not know. It was of medium size, and seemed to get more tame as time went on, but when frightened or angry it acted always very much as a perfectly wild specimen. The marmot is a creature of low intelligence, a natural result, perhaps, of so much of its life being spent in a burrow, and so much of its time being drowsed away, free from that struggle for existence which has apparently been so great a factor in all kinds of development.

On account of the destructive power of its teeth it became necessary to have a cage constructed of strong wire, with a suitable movable bottom, essential for cleanliness.

My specimen would eat fruits, such roots as turnips and carrots, bread, biscuits, etc. But he became very fond of porridge with a little milk, and when he was disposing of this, his smacking could be heard all over the house. It seemed to agree with him perfectly.

The object of my investigations being to ascertain not only the nature of the winter sleep, but the character of its variations under a changing environment, especially as regards temperature and meteorological conditions generally, I kept the animal in different rooms of the houses in which I lived successively during the creature's lifetime, and also in different parts of an outhouse in which my dogs, fowls, etc., lived.

It would take up too much space and prove rather wearisome to attempt to reproduce here the records which I have kept in detail. It will be both more practicable and more profitable to give the general results of studies on this one specimen for the four years during which he lived, and then refer to the unexpected result of the study of another specimen during the past winter.

I think the following life-periods were recognisable in the specimen I had under observation:

(1) A period characterised by either drowsiness or sleep or profound torpor, lasting from about November to April.

(2) A period of perfect wakefulness immediately following, during which the animal was emaciated, in

poor coat, and with a general low vital capital, lasting for some weeks.

(3) A period of improving condition, with good health and a desire to get free, which latter was also noticeable in the preceding period.

(4) A period of maximum weight and vigour, with perfect coat and an abundance of fatty tissue.

But little need be said about any of these life-periods except the first. During the second the emaciation increased rapidly at first on waking, and was equally marked by voracious feeding. The meaning of this will be referred to later. It is plain that the organism entered on its long period of diminished vitality with a large stock of reserve material, and it is equally clear that this was drawn upon to the full.

I now turn to the peculiarities of the sleeping or torpid condition. I have characterised the first period as one of drowsiness, or sleep, or torpor, because there are sub-divisions of the first period during which the animal was found in a condition that was characterised by drowsiness and no more; again it was plainly only sleeping, while again it was as profoundly torpid as it ever became. The period of most profound sleep was never reached all of a sudden, but was preceded by the two states referred to above. Moreover, as the depth of winter approached the sleep became more profound, and the reverse with the approach of spring, so that we might represent the depth of the sleep by a rising followed by a falling curve with a rather long, flattened top. During the whole of the first period the animal nestled in the straw, with which he was always provided, and when he was most profoundly unconscious but little of him could be seen, often so completely was he covered.

Another important matter: The amount of food consumed was directly proportional to the depth of his sleep. Not only did he eat less frequently, as a matter of course, but the quantity taken at one time was less. Occasionally, when fully hibernating, he would awake to fall asleep again, merely after gathering the straw about him a little closer, and not eat at all. His cage always being supplied with food, there was no doubt about his ability to provide himself when so inclined.

His excretions were also in proportion to the amount of food consumed, and especially to the depth of the sleep. The less he ate, and particularly the more profoundly he slept, the less were his evacuations when he awoke. He never befouled his bed, but always left it to void urine and fæces.

During the period of mere drowsiness the animal would be awakened by a gentle rustling against its cage, and even when asleep, a noise, etc., would always arouse it, but when in a torpid condition it could not be thus aroused, but might be handled without being brought to the natural condition, though handling and much less disturbance always caused movement, a phenomenon to be discussed later.

In the spring of 1891 my Woodchuck came out of his winter sleep in a very emaciated condition, and this, as usual, increasing after his awakening, he was in a doubtful state; but the case was soon decided against the animal in consequence of my servant having left him for some time fully exposed to the sun's rays. An autopsy revealed the fact that the animal (a male) was the subject of tuberculosis of the lungs, though possibly but for this exposure he might have lasted another year.

Through the kindness of Mr R. F. Rorke, an undergraduate in medicine of M'Gill University, I became possessed of two specimens of the Marmot in the latter part of the summer of 1891. One of these was very large and in fine condition, and his escape soon after arrival has been a frequent subject of regret by me, as I hoped to be able not only to continue the observations, but to make them comparative, as it was my intention to keep both under the same conditions—in fact, in the same cage. But the curious part remains to be told. Though I kept the remaining Woodchuck under exactly the same conditions as the animal I had had so long, he did not hibernate for an hour the whole winter, though he drowsed and slept enough.

It has been generally considered that the hibernating conditions of animals was dependent above all else on the temperature of the surrounding atmosphere. My experiments in bats seemed to warrant this conclusion, for whenever the temperature reached the neighbourhood of 45° F. to 40° F., the other conditions being favourable, my specimens began to hibernate. It was also true that my Woodchuck was in the deepest sleep during mid-winter when the cold is greatest. Whether a bat could be put into a state of torpor during summer by exposing it to a lowered temperature, I do not know. However, Marshall Hall maintained that the diurnal sleep of the bat ("diurnation") was exactly the same phenomenon as the winter sleep. The same writer maintained that hibernating bats always awoke when the temperature fell below freezing point, and his observations showed that the temperature of the animals was always a few degrees above that of the surrounding atmosphere. Probably Hall is correct in the main, for my bats, when the temperature sank during the night much below freezing, were always found dead in the morning. Whether they awoke first, or simply passed from torpor to death, I do not know.

However, for the Marmot, I can assert positively that this rule does not hold, for frequently the water was found frozen in the apartment in which the animal was kept, yet he was undisturbed. Nevertheless, I came to the conclusion that this exposure is injurious to a hibernating animal, and that it had something to do with the poor condition in which my specimen was found in the spring of the year in which he died.

Before discussing the true nature of the phenomena of the winter sleep, I call attention to certain cases of allied nature.

Such frequent references as we find to the hibernation of swallows deserve some consideration.

It is also stated that in Scotland sheep have been found alive after being buried for weeks beneath the snow; and Dr Frank Miller of Burlington, Vt., reported, when a student, to the Society for the Study of Comparative Psychology at Montreal, that hogs had been found alive after being accidentally imprisoned below the surface for several weeks longer than it seemed possible for them to survive under ordinary circumstances, so that it would appear they had been in a condition of hibernation or some such state.

Turning to the human subject: We have all read of instances in man of "suspended animation," or "trance." The case of Fakirs in India having been buried alive, exhumed, and resuscitated after months, is attested by such evidence as it is difficult to set aside, however hard to credit.

Mr D. W. Ross, a student in medicine of M'Gill University, has gathered the facts of a peculiar case in so far as they are now obtainable. The individual in question was known as "Sleepy Joe," a farmer by occupation. He was married and had several children, one of whom, a girl, had the same drowsy appearance as her father. This man would sleep almost constantly for several weeks, awakening, however, to attend to Nature's calls and to take food. He would at times awake more fully and then set to work, whether it was day or night, and almost incessantly labour as if to make up for lost time. He was rather weak mentally, but appeared ashamed of his sleepy tendencies, which seemed to get worse as he grew older. He lived to be about sixty years of age.

Dr Aug. Robinson of Annapolis has kindly given the following notes of a unique case:

"John T——, son of a pensioner, is now about sixty-two years old. When he was twenty-eight years of age his father committed suicide by cutting his throat in a fit of insanity. Before his father's death John had shown symptoms of melancholia. He would sit by the hour over his father's bench (cobbler's) laughing and talking to himself, and working himself into a frenzy, fighting imaginary foes, and going into immoderate fits of laughter.

"I cannot ascertain, after much enquiry, how long this condition of things lasted before he lapsed into his present state, but this much is certain, for the last thirty years or more, about the end of September every year, he falls into a deep sleep or stupor, and, as his present condition is a fair sample of the manner in which his winters have been passed since he was first attacked, I will describe him as I found him on Monday, 10th December 1888, and repeat what I was told by his friends regarding his attack this autumn:

"About 31st August Jack went to bed after eating his evening meal, as usual, without exhibiting anything out of the common in his manner or otherwise, or giving any reason for the supposition that he was out of sorts in any way. On the following morning he did not get up, nor has he shown any more vitality than any sleeping man up to this time. His sleep is very quiet without any stertor, indeed it is as calm as that of a child. Twice in every twenty-four hours he is taken up, a person supporting him on each side, holding a vessel for his convenience. He knows enough to voluntarily empty his bladder. The urine is high in colour and scant in quantity. About eleven o'clock every night he seems to show rather more life than at any other time, and advantage is taken of this to pour a little thin oatmeal gruel, beef-tea, or soup down his throat, he opening his lips to allow them to do so, and slowly swallowing it. He only takes a very little each time, and, if urged to take more, simply keeps his mouth shut. About once in every thirty days, not exactly at regular intervals, during the evening generally, the family will hear a peculiar chattering noise. They never take any notice of it, for they know it is Jack going down to the outhouse to empty his bowels. He then returns to his bed and sleeps. He knows enough to throw a quilt over his shoulders at such times.

"At the time of my visit I found his temperature 96° F., pulse 60, regular, though not strong; respirations 14, easy and quiet, skin cool. A pin stuck into his arm caused no apparent change, and he might be pinched until black and blue without its causing him the slightest uneasiness.

"My first visit to Jack was about twenty years ago, when I first came to live and practise in the vicinity, and it came about in this way. Of course there was a talk about the new doctor, and what he could do, so I was called to see this queer case. I got all the particulars from the friends and neighbours, and what means had been tried by other doctors, and then I promised to try what I could do. On the following day I went again, accompanied by my brother, also a physician. We took with us a good galvanic battery. One of the handles was placed in each hand and bound closely to the fingers with wet bandages. We then put on the full power of the instrument. Poor old Jack was out of bed in an instant, and I shall never forget his look of astonishment and horror as he yelled out: 'Damnation! What's that?' I can also well remember my own feelings of satisfaction and complacency when the 'natives' congratulated me on my success in this, my first case. I walked off, as if saying to myself: 'I knew I could do it.' Well, Jack remained awake about three days, and then I got a message that my patient was off again. I went up and tried the battery a second time, with only the effect, however, of making him open his eyes and grunt out "Eh?" in a querulous manner, and after looking about him for a half hour or so, he lapsed into his former condition. Next day I again tried the battery but without the slightest effect, so I gave it up as a hard case.

"This is all I have to say about this singular object, except that, of course, he becomes very thin and haggard before he rouses finally in the spring, and he does not fairly waken until the end of May or first of June. During the summer months Jack does exactly the work he is told to do, but he must be told over again every day; for example, if desired to bring the cows from pasture he will do so, but will not milk them until told to, nor will he turn them a-field again without being desired. He does not seem to know what to do next, even though the same routine is carried out every day. An exception, however, must be made in one respect. He does not require to be told when dinner or tea-time arrives, and is blessed with an excellent appetite. Jack is always ready for his food, and is not particular about quality, so that quantity is there. He will talk quite rationally on any subject when spoken to, and recollects distinctly most of the incidents of his childhood. He will hold animated confab with the cows, dogs, trees, wheelbarrow, or any other object which happens to be in his way, and may be noticed sometimes lecturing a tree for some time, breaking out occasionally into uproarious fits of laughter."

When visiting, in 1890, Dr C. K. Clarke, Superintendent of the Asylum for Insane in Kingston, he happened to make reference to a peculiar individual known to a few as "the sleeping woman."

It at once occurred to me that her case would prove a study of great value if it could be carried out, and I suggested to Dr Clarke that he endeavour to supplement such facts as we could learn, and such observations as we were able to make by a joint visit to the subject of this peculiar condition by a careful study of the case. He succeeded much beyond my expectations in doing this, and has published the results of his investigations in the *American Journal of Insanity* for October 1891, under the title, "A Case of Lethargy," and from that paper I extract the principal facts in regard to this case, which is so remarkable that it may be well to state them somewhat fully.

"Several seasons ago I heard that there was a patient who had been in a trance for years, and from time to time word was brought to the effect that the condition still persisted, and that all efforts to rouse the woman were without result.

"A little more than a year ago I obtained permission to visit the patient, but was not allowed to make any extended examination.

"On entering the room I found a thin, old woman in bed, apparently fast asleep. Her respirations were irregular and varied much during the visit (lasting almost half an hour), running all the way from 24 to 44 per minute. The pulse quickened in a marked way during my stay, and ran up from about 80 to 120. The woman had her eyes half closed, and to all appearances was oblivious to everything that was going on.

"The nurse gave many details regarding the patient,

and made a number of statements, some of which I shall repeat in a few minutes. Many of these statements we were able to verify at a subsequent period; others were undoubtedly inaccurate. Before going into details regarding the every-day life of this case of lethargy as we saw it, perhaps it would be well to give a brief outline of the history of the patient. Unfortunately, it has not been found possible to get as many important facts as could be desired in connection with this history, but great care has been taken to eliminate all doubtful points. The patient was born in 1820 or 1821, and when she came under observation was almost sixty-nine years of age. The neurotic history was pronounced on 'both sides of the house'; evidence going to show that her father had suffered from melancholia. A reliable person states that the father died from 'softening of the brain'—possibly general paresis. The patient's mother was subject (a member of the family states) to attacks of partial loss of reason, which could only be cured by change of air and surroundings. It has not been possible to get an accurate account of these attacks of 'partial loss of reason.' The patient's early history is not well known, but it has been stated on good authority that she was "peculiar," and in childhood complained of some head trouble that caused her to keep her hair cropped short. She was married when very young, probably when seventeen or eighteen years of age, as she was but twenty-one when her third child was born. Three years after the birth of her last child she was noticed to undergo a change in disposition and acted 'strangely.' She could not be depended on, was untruthful and whimsical, and worried a great deal about trifles."

It is from a subsequent period (three years later) that the history of this case must be dated.

The son, the youngest child, says: "The first recollection that I have of mother's sickness was when I was six years old. My little sister had died, and I was just recovering from an attack of scarlet fever when she was taken down. I think the attack was caused by grief over sister's death, and over-exertion and want of rest. I do not remember how long she was sick at that time, but I recollect that her left side was completely paralysed, and that after a time a strong liniment was used which partially removed the paralysis, and when she went to the country for several weeks she came back well.

"The second time that she was taken sick was when I was twelve years old. She felt poorly for some time, and was then taken with fits, or convulsions, which lasted for, I think, three days, having sometimes four or five in an hour. She was confined to her bed for, I think, about two years, in very much the same condition as at present. I do not know what curative means were employed, but she gradually improved a little, and was again sent out into the country, where she seemed to recover.

"She enjoyed pretty good health for about six years, but had to be very careful; she never drank tea or coffee, and always had to have the hair on the back of her head cut short.

"About this time her father was taken sick, and we moved into his house to take care of him. This seemed to affect mother, and after a short time she was again taken with fits, and soon went into her former condition.

"During this sickness, which lasted about two years, she used to sit up a great part of the time, and appeared to be perfectly conscious. She knew father and those who waited on her, used to call me her boy, but appeared to be in a sort of stupor most of the time. She was again sent into the country and came back well.

"Then ensued a brief interval of about two years, during which time she was in fair health, but was again taken down as before, and was sick for nearly seven years. During part of this time she was very low, and we watched day after day at her bedside, expecting each day would be the last; but she again rallied, and gradually her bodily strength and reason returned to her.

"She was well for about five years, when she seemed to be taken with a low fever and gradually went down to her present condition."

Such is the son's account of the case, and from it we learn that the history of nervous trouble dates back at least forty years, and the inference is that the first indications of lethargy then made their appearance.

The details of the conditions that existed during the different attacks are almost entirely wanting, and it is unfortunate that we are left almost completely in the dark regarding the nature of the convulsive seizures that occurred. Subsequent history would lead us to believe that these were, in all probability, hystero-epileptic in origin.

About the year 1862 the patient fell into a state of lethargy that lasted for seven years or more. The condition was not one of complete unconsciousness, and although the woman appeared to sleep almost continually, occasionally she would wake up for a minute or so at a time, and converse in a rational manner.

It is not possible to make more than general statements in regard to these attacks, but it is beyond doubt that the conditions were not identical with those that characterised the last attack. Evidently the condition of lethargy was not so profound.

The announcement of the death of a warm friend was the immediate cause of her awakening. The return to even an approach to a normal condition of health was a very gradual process. She was lachrymose and childish for some time, and could not use her limbs properly for months—in fact, had to learn to walk again.

During the period of wakefulness that now ensued—seven years or more—the patient, to a certain extent, interested herself in the affairs of every-day life. She went about the house, etc., but was very quiet and did not seem able to concentrate her mind on anything. Her memory was markedly deficient, and she seemed astonished to find people and places changed, and could not realise the fact that she had been asleep for such a long time.

When waking up from her long sleep, one of the first requests made was for beer, and strange to say, the same want was expressed many years after when arousing from a subsequent attack.

About thirteen years ago the patient gradually passed into the condition in which we saw her. At first she spoke occasionally, but in a childish manner, and often made a request for meat and potatoes, invariably using the following words: "Meat and potatoes, a plate all full up to the top!"

Before giving the details of the case as we saw it, it will be well to repeat, in a general way, the statements made by the nurses who had the care of the patient before she came to the asylum: She seems to exercise a certain amount of discrimination regarding her food, and will eat enormously or not at all, and when her appetite is not lost, does not seem to know when she has had enough. Her diet is made up of minced meat, potatoes, soft toast, milk, etc., and she is particularly fond of meat and potatoes—in fact, will not touch anything until meat and potatoes are provided. She does not like sweet things. When not suffering from diarrhœa, eats three times a day. Eats as much as any healthy, active woman of her age. Objects to nauseous drugs, and endeavours to push the spoon away with her left hand.

The attitude during the day is quite different from that assumed at night, and the patient undoubtedly sleeps more soundly at night than during the day. In the daytime her legs are extended, at night, drawn up. In the daytime she is put either on her back or right side, at night on her left side, and remains in this position until morning without moving—in fact, cannot roll over. She will not settle down for the night until a drink of cold water is given. In the daytime, sometimes for an hour or so at a time, appears to be nearer a condition of consciousness than at any other time. This occurs generally after breakfast, but she has to be roused for her meals.

When heavy coverlets are put on the bed she attempts to shove the blankets off with her left hand, and likes to be very lightly covered. The eyes are three parts closed during the day and completely closed at night. The face sometimes becomes flushed. She never speaks, and, in fact, has spoken but once in eleven years or more, and that was quite recently (1890), when she said: "I am not asleep." Her appetite has been better since she has been in the long sleep than it was before, and she eats things she would not touch when awake. At least once during the present attack she has, unassisted, got out of bed, and there is reason to believe she has done the same thing several times, but not within three years, as her physical condition renders it impossible. Several times the nurse fancied the patient was moving about the room at night, but for some time could not actually prove that such was the case. At last, however, a fall was heard in the middle of the night, and the patient was found lying fast asleep at the bottom of the stairs, down which she had fallen.

During the present attack she has fasted on several occasions, and once went fifteen days without food.

It must be remembered that the nurses were speaking of the last attack, and at a time when the patient had been in a state of lethargy for more than eleven years.

In September 1890, I saw the patient with Dr Clarke. This was my first, but Dr Clarke's second visit. We found the patient, an old woman, in bed. She was lying on her back with her eyes half closed. Her face, when we first entered, was somewhat flushed and respiration rapid. When respirations were closely observed it was noticed that they were most irregular, and at times ceased for several moments. They averaged 22 per minute. Pulse was 104, fairly strong and regular, arteries almost free from rigidity. Axillary temperature 98⅜°. The nurse stated that ordinarily the patient's bowels moved but once in three days, but latterly she had developed a tendency to diarrhœa, and since that had evinced a sense of discomfort until the bowels were relieved. This sense of discomfort was evinced by whining like a dog. Ordinarily she would not give any indication that she wished to relieve herself, but the presence of the bed-pan would excite the reflexes. She does not soil the bed. The statement of the nurse in regard to the amount of urine passed every day was, that a little more than half a pint would be a fair average.

A physical examination of the patient was made. The left foot was drawn as if there were a contracted Tendo-Achillis; right foot drawn down but not in such a marked manner as the left. Marked rigidity of the right knee and leg; left leg and knee not rigid. Right ankle easily moved; left rigid. Patellar reflexes absent. Tickling the soles of the feet did not cause any evidence of sensibility. Each great toe was drawn under the second toe, this condition being especially

G

marked in the left foot. When the soles of the feet were tickled it was thought that the respirations were slightly deepened, but on account of the irregularity of the breathing it was difficult to determine this point, and it was considered undecided. Patient's hair grey; nails healthy and not abnormally brittle. Facial reflexes better than reflexes in any other part of the body. Orbicular reflexes good even with air; at the same time it was noticed that flies crawling over the face did not excite the reflexes. Pupils responsive to light. Small bedsores found on hips, and evidence of former deep-seated bedsores plainly visible.

While we were present the nurse endeavoured to arouse the patient, and tried to get her to take some food. A feeble protest was made (whining), the patient winked for a few moments, and then went off to sleep again. Bread was put in her mouth, but remained there without any effort being made to swallow.

On 9th October 1890, the patient came under Dr Clarke's care, and a series of observations of the most complete character was made.

The patient, a thin old woman, apparently not weighing more than sixty pounds, was carried into the infirmary from the ambulance, and placed in bed. She was asleep, and did not seem disturbed by the jolting to which she had been subjected. Her temperature was 97⅜°, pulse 107, and respirations 20. Efforts were made to arouse her, but without avail. Friends stated that she had been in her present state of lethargy for more than eleven years.

Her eyes were half closed, and it was found almost impossible to get her to swallow anything. Next morning her temperature was about normal; pulse 117, respiration 18; still asleep with the eyes half closed, as she remained nearly the whole time she lived.

She was under observation from October 1890 until February 1891, when she died. In these four months she was closely watched, and until the last week of her life gave little indication that she had the slightest knowledge of the fact that she lived.

She would remain in any position in which she was placed in bed, and if not fed would undoubtedly have died without making any sign that she required or desired food. Her temperature was almost invariably sub-normal, sometimes falling to 95°, although occasionally it would rise to nearly 102° without any cause that could be determined. Her appetite was capricious, although she undoubtedly had decided likes and dislikes in regard to food. She preferred beef and potatoes to anything else. The process of eating was very slow, and sometimes it would be more than an hour before she could finish a meal. When she drank anything, milk was evidently preferred. She was very clean in her personal habits, and never soiled the bed.

The quantity of urine passed was very small, not averaging more than one-seventh of normal. The bowels moved but seldom, sometimes only once in six or seven days. It was possible to rouse her for a moment or so to the extent of making her open her eyes, but beyond this she would give no indication of consciousness, and went to sleep again immediately. Her legs were nearly always drawn up, although when the patient was admitted it was stated that she always straightened her legs at night. Her feet were almost invariably very cold, and the hands sometimes so. Occasionally her eyelids would tremble and quiver, just as they will in a patient suffering from hysteria. Generally, when much bothered, she would for a few moments make a sort of whining protest.

The facial expression was quiet, almost death-like, under ordinary circumstances, but sometimes, when undergoing examination, an expression suggestive of pain would appear; at the same time it is questionable if pain was really experienced, as the heart and respirations did not show the least disturbance.

A large amount of food for one so frail was consumed in a day, although on some occasions the appetite was completely lost. Sometimes, when suddenly disturbed, she would start nervously, and her hands would tremble. Trained Nurse Osborne, who was with her very constantly, seemed to think that there were times when she was nearer a condition of consciousness than at others, and as this statement was also made by the former nurses, possibly it is correct.

Occasionally she would push down the bedclothes with her hands, and the history of the case would go to show that heavy coverings were always objectionable.

Nearly every day she was propped up in a chair for half an hour. This did not seem to have any effect on her general condition.

The account of her last days are interesting: Early in February 1891, a marked change took place in the patient's condition. Diarrhœa developed, and the woman was evidently suffering pain. On the 4th of February she was undoubtedly awake, and in the evening spoke in a hoarse whisper, asking for a sour drink. This was the second time she had spoken in thirteen years. On the morning of the 5th of February again asked for a drink, yawned twice, and fell asleep again. In the afternoon was again awake, fed herself in an awkward way, and in the evening spoke again in a natural manner. I sent for her friends, and they endeavoured to get her to take notice of them, but she did not appear to know them, and went to sleep as usual. The trained nurse's notes for the next few days are as follows:—

February 6*th.*—Will feed herself with bread or anything dry. Hand shakes too much to use a cup or spoon. Will ask for anything she wants, but will not speak at any other time. Always uses her left hand.

February 7*th.*—I was called in about 4 A.M., and found her lying on the floor; she would not speak, but from all appearance no one had touched her; she had evidently got out of bed herself. At 9.30 A.M., she was cold and very white looking; about fifteen minutes later her face was flushed and moist, body warm, hands, knees, and feet cold. This soon passed off, leaving her in her former condition. Temperature was $95\tfrac{2}{5}°$ lower than at 8 A.M. This afternoon asked for a sour drink and a big cake. Spoke hurriedly, but quite loudly and distinctly. Kissed the nurse twice when asked to do so.

February 9*th.*—Has not been well at all to-day. Moaned when disturbed. Has eaten scarcely anything, but has taken more milk than usual. Has had slight diarrhœa since last Tuesday; worse to-day.

February 11*th.*—Asked frequently for drinks to-day, and last night said her throat was burning. Does not appear to recognise any of her friends, nor to realise that she is among strangers. So long as her wants are attended to she seems quite unconscious of anything else — not exactly unconscious either, but as though she took no interest in what went on around her.

February 12*th.*—Is better this morning; had no diarrhœa during the night. Has asked three times for something to eat, which sounds like meat, but when I get it for her she won't eat it.

February 13*th.*—Diarrhœa much worse to-day.

February 15*th.*—Diarrhœa somewhat better.

On the 16th she was slightly better, and asked for beer and cocoa, and said she felt as if she were burning up. From this time she steadily grew worse, and died on the 26th.

Dr Ruttan, Professor of Chemistry, M'Gill University, made elaborate analyses of urine sent to Montreal from time to time, and without offering any detailed statement here, I may say that the general conclusions arrived at were as follows: The whole of the urine passed in six days was sent, and he says the total amount, if representing six days' urine, is about one-seventh the normal. This contains all constituents in about normal quantities in relation to the volume of the urine, except the phosphoric acid, which is about one-third what it should be.

I.—AUTOPSY.

Inspectio Cadaveri.

Nutrition poor; body much emaciated; apparent age 65 to 70; weight about 50; rigor mortis complete. A. M. staining on hands and feet; P. M. staining on back of trunk; bedsores on sacrum, tip and ball of great toe; feet and ankles œdematous; legs flexed on thighs by contracted tendons; no teeth, and sockets much absorbed.

Sectio Cadaveri.

HEAD.—Scalp thin and easily dissected; calvarium of average thickness; tables thin, however, diploe being in excess; dura mater not adherent to the skull, slightly opaque at vertex; one slight adhesion to brain at margin of longitudinal fissure; ante-mortem clots in longitudinal and lateral sinuses, the clots in the lateral sinuses being particularly well organised.

BRAIN.—The brain weighed about 35 oz.; microscopically, it was healthy in appearance—in fact, in asylum experience, I have never seen as healthy a brain in the post-mortem room.

Convolutions well marked and sulci deep; grey matter abundant; brain substance firm; ventricles free from evidence of disease; brain not examined microscopically.

THORAX.—Sub-sternal adhesions. Emphysema of cellular tissue beneath sternum; cartilages not ossified.

HEART.—Small—weight $3\frac{3}{4}$ oz. Pericardial fluid in average quantity; blood in great veins, and right auricle fluid; walls of right auricle and ventricle unusually thin; valves normal; small post-mortem clot in left ventricle; walls of left ventricle hypertrophied; left auricle normal; valves of left side normal.

AORTA.—Ascending aorta dilated into a fusiform aneurism; capacity about twice that of normal; arterial coats not thinner than normal; no evidence of atheroma; no pressure effects noticed; varicose veins on posterior walls of the heart; abdominal aorta atheromatous; ante-mortem clots in abundance.

LUNGS.—Right: Very adherent at apex; small adhesions all over surface of lung; apex, a mass of tubercle—in fact, tubercles were found scattered throughout the whole of the lungs, and in the apex a small cavity existed; hypostatic congestion marked. Left: In this lung a certain amount of hypostatic congestion was apparent, and an occasional tubercle was found, otherwise the lung was normal; cord-like adhesions of pleura on surface.

ABDOMEN.—Liver adherent to chest walls and diaphragm; whole capsule torn off in taking out, and remained attached to diaphragm and abdominal wall; weight, 20 oz.; three vertical furrows present on anterior surface of right lobe; these furrows were about 2 inches in length, centre one distinctly marked; nutmeg condition present.

STOMACH.—Large; about 2 inches from pyloric orifice was a constricted portion.

This condition was undoubtedly not the result of any inflammatory action, but the natural shape of the stomach, giving rise to an appearance suggestive of a rudimentary second stomach.

INTESTINES.—Small; evidences of an old peritonitis; adhesions everywhere; there were several constricted portions from 3 to 6 inches long; in no place was there complete stricture, and no scars were present; above the constricted portions of the intestine was much distended.

CÆCUM.—Walls much thickened and much venous congestion; inner surface dark red and roughened; had appearance of numerous varicose veins in wall.

ASCENDING COLON.—One portion constricted, and part preceding dilated; transverse colon, normal; descending, slightly dilated.

KIDNEYS.—Right: very small, about $2\frac{1}{4}$ inches long; apparently normal. Left: about 1 inch longer than right; apparently normal; capsules non-adherent.

We may sum up the case by saying that, in the subject under consideration, we have a woman inheriting from parents an intensely neurotic organisation, in consequence of which she showed many indications of an ill-balanced and unstable nature, culminating in various vital crises, including periods of stupor. In fact, this woman seems to have spent nearly one-third of her whole existence in an unconscious condition, being then a purely vegetative organism.

At one period of her life she was a veritable Rip Van Winkle, finally sinking into a long lethargy from which there was only a brief consciousness prior to the final stoppage of the vital mechanism. But it is to be noted that this curious condition was not the result of any gross lesion of the brain, but of hidden molecular peculiarities, which renders the case, to my mind, all the more instructive when considered in connection with all those states I am now considering.

II.

I propose now to discuss the real nature of hibernation and kindred states.

In the paper on Squirrels read before the Society in 1887, I said, speaking of hibernation: "I think it is very probable that, when the matter has been fully investigated, all degrees of cessation of functional activity will be found represented, from the daily normal sleep of man and other animals to the lowest degree of activity consistent with the actual maintenance of life."

As a matter of fact this is the conclusion toward which all my investigations since that time have tended. Though some maintain that in true hibernation there is cessation of respiration, it would be hard to prove this, for, as Hall showed, the circulation continues, and the very beating of the heart against the lungs displaces a certain amount of air, and in any event we cannot leave out of account diffusion of gases, which, in all cases of animals with lungs, plays an essential part in the process of respiration.

It would be interesting to know the condition of the heart in a hibernating frog or turtle; but in such creatures the skin, as also probably in snakes, has a respiratory function. Live frogs will stay for hours at the bottom of a tank in winter, provided fresh water is flowing over them constantly. In fact, winter frogs kept under these conditions respire largely by the skin. So far as the bat is concerned, it is difficult to observe any respiratory movements; but in the Woodchuck I never fail to notice them at considerable intervals, say once in five to nine minutes, even when hibernating most profoundly. The respirations were peculiar. Sometimes one deep inspiration, preceded by a sudden relaxation of the enlarged chest, would be succeeded by a long pause; again there would be a series of very slight respiratory movements. It was always possible by the respirations alone to predict when the animal was approaching the waking state.

The awakening was never sudden but gradual, often extending over hours. I have seen something like this, though less remarkable, in the bat. This is no doubt protective to the vital machinery, for when Hall maintained that bats, suddenly awakened from the hibernating condition, died speedily, he was, in the main, if not entirely, correct.

A brief consideration of natural daily sleep will throw light on winter sleep, etc.

Sleep is favoured by moderate exhaustion, a good condition of nutrition, and the absence of all sorts of peripheral stimuli.

The case of the youth having but one good eye and one hearing ear, who could be put to sleep by closing these up, is very instructive. This lad did not, however, continue to sleep indefinitely, but awoke after a number of hours, showing that though there are certain conditions in the environment that favour sleep, the latter is essentially a condition of the central nervous system, and dependent on laws governing the latter. This view makes it clear that sleep is naturally a nocturnal condition for most animals, owing, no doubt, to the evolution of life in relation always to the environment. The fact is, we cannot conceive of life except in, and by reason of, in a sense, some environment. The change of the seasons, day and night, and all the periodicities of the inorganic world have, as a natural consequence, stamped themselves on living things, plants as well as animals. Sleep, then, is essentially a rhythmic diminution of the activities of all the tissues, but especially of that one which controls all others, the nervous. Rhythm seems to be at the basis of all things organic and inorganic, but has not been enough considered in our explanation of living cells. It was long the custom to explain sleep by anæmia of the brain, whereas the very anæmia was due to a contraction of the blood-vessels of the part, accompanied by the diminution of the heart-beat, both of which are periodical and dependent on the rhythm of the nervous system itself. Of course, temporary anæmia of the brain favours sleep, though it is not the essential cause. As a natural consequence of the decline in the activity of the great controller of the cell activities (metabolism), *i.e.* the nervous centres, it is found that all the functions of the body, without exception perhaps, are diminished during sleep. Marshall Hall and others since his time have shown that the gaseous interchange in a hibernating animal is greatly lessened. This diminished metabolism explains why the animal does not require to eat, or but little. It explains the diminished excretions, etc., etc.

This being understood, it is not surprising that hibernating animals may be kept under water for long periods as is the case with newly-born mammals, as kittens and puppies, whose vital machinery as yet works very slowly, which are, in fact, in a condition but slightly more advanced physiologically than the uterine, which is a sort of reptilian pre-natal state, as regards the circulation, respiration, etc.

As the metabolism of reptiles and amphibians is of a much slower kind than that of mammals, it is not surprising that their winter sleep is more profound; but it is to be observed that the change from their most active to their most sluggish condition is not probably relatively greater than in the case of mammals.

All forms of profound winter (or summer) sleep are protective, both of the individual and the species. Manifestly amphibia, reptilia, and other groups of the animal kingdom must have utterly vanished from the face of the earth but for such a power to adapt to conditions. Probably many individuals, if not some entire groups, have, through more or less complete failure to adapt, disappeared before this habit of the nervous system and of the whole organism became perfect enough.

It is equally clear from the investigation given to the subject that hibernation, like daily sleep, is not a fixed and rigid thing, but just as it has been the result of adaptation to the environment, by virtue of the plasticity of function of all living cells, so the power to modify still remains.

It is possible to conceive of its being lost in certain groups of animals—indeed this phase of the subject has been as much impressed on me as the other. Sleep, hibernation, and all such states are not invariable, but to a certain extent, so dependent on the surroundings that—as in the case of my last marmot, also of turtles and frogs kept within doors—there may be an omission of that condition which is habitual under the normal environment of the animal.

I would like to emphasise these facts, for they seem to me to throw great light on the evolution of function at all events, and on those changes which may become so great as to lead, we can hardly say to what, in the lapse of time.

For years I have had turtles, and especially frogs, under observation during the winter months. Our frogs for laboratory use at M'Gill University are kept in a tank in which the water is being continually renewed by a slow stream. They are not fed. None of the frogs seem to pass into a condition of true hibernation, but they descend to the bottom of the tank and remain quiet, as if asleep or partially torpid, as, indeed, I know they often are for hours. In this is an interesting modification of that most profound torpor which they experience when buried in the mud of ponds.

Even in the winter life of a creature like the marmot we may have all degrees of drowsiness or torpor, as I have shown, and it is not to be forgotten that our own daily sleep has its degrees, so that the night's sleep may be represented by a curve with a sharp rise and very gradual fall, which may, as we all know, be greatly modified by circumstances.

The same laws seem to apply to all the known cases of human lethargy, hibernation, sleep, or whatever the state may be called. In the case of the buried sheep and hogs the protective value of the condition is evident, as also in the case of the lethargic woman. This individual, with so ill-balanced and unstaple a constitution, would probably have been carried off by some form of actual disease long before, had she remained awake. She could exist as a mere vegetative organism, but not as a normal human being in the ordinary struggle for existence. One thing which has been much impressed upon me by my studies of this whole subject, is the varying degrees of sensitiveness to temperature and meteorological conditions in different groups of animals and different individuals of the same group. The bat as compared with the marmot, for example, may be worked like a machine by varying the temperature. On the contrary, the degree to which the woodchuck is independent of temperature was a surprise to me after my experience with the bat. But the woodchuck answered like a barometer to predict storms. In fact, I am satisfied that many wild animals have a delicate perception of meteorological conditions which man has not, and which, in a sense, makes them wiser than our science, and wiser than they know, for they act reflexly, as it were. Often my marmot would be heard in the night scraping the straw about him prior to a storm that did not reach us for many hours after.

Marshall Hall laid it down as one of his principal conclusions that in hibernating animals "muscular irritability" is increased.

If the term reflex be substituted for muscular, I believe the conclusion is correct. I found, as a result of scores of trials, that when the marmot was hibernating, he was more sensitive to slight stimuli, such as blowing on the hairs of the skin, than when merely sleeping. Plainly this was not a case of muscular irritability at all, but it does indicate that the reflex mechanism is more excitable, as it is, for example, in an animal under the influence of strychnine, and as it is in animals from which a portion of the cerebrum has been removed.

It may be because the unconsciousness is so profound, *i.e.* the brain so far from its ordinary functional activity, for it is well-established that the brain inhibits the spinal cord normally, to a certain extent.

Apparently this increased reflex excitability must be to the advantage of a hibernating animal, for the cord and medulla oblongata are the parts of the nervous centres that especially preside over the functions of organic life, which are necessary to maintain a mere animal existence.

All problems of a biological kind must ultimately be referred to cells, and so with this of hibernation. Indeed, it would seem that unicellular animals pass into a condition which is related to that of hibernation. The so-called encysted stage of protozoa is perhaps analogous and similarly preservative of the individual and the species.

The study of a subject like the present one gives rise to many questions. Can the molecular machinery of life entirely stop, and yet be set in motion again? We know that cold-blooded animals may be frozen and completely restored to a natural condition. This and the encysted condition of protozoa are suggestive of such a possibility.

Yet in insects a condition of perfect quiescence is accompanied by the most wonderful changes. The worm-like caterpillar becomes within its cocoon the butterfly, with locomotive powers immeasurably greater.

For myself, the more I study biological problems, the less am I inclined to subscribe to rigid formulæ of being. The study of a single group of animals from a physiological point of view, much less that of a single individual, does not suffice to enable one to lay down laws that will apply to similar processes in other groups of animals except in the most tentative way. I can never forget the lesson of my marmot that did not hibernate at all, and what modification of present views more extended study of this subject of sleep in all its phases will produce, it is impossible to say.

All my own studies have greatly impressed me with the plasticity of living things, their power to adapt to altered environments, and, if I might suggest one of the great changes that is likely to come over the biology of the future, it is a recognition of the above fact; so that we will cease to generalise so widely from such narrow data, or rather, perhaps, we will be ready to believe that phenomena, very different from those we know, may be possible in the realm of living things.

It has often occurred to me that possibly, if such consideration were kept in mind, we should not be so disposed to assume that the conditions under which life now exists are precisely those under which it always did exist.

PART III.

THE PSYCHIC DEVELOPMENT OF YOUNG ANIMALS AND ITS PHYSICAL CORRELATION.

I.—The Dog.

Introduction.

For mind and body alike the past determines the present in no small degree; hence it follows that the more perfectly the history of each step in the development of mind is traced, the better will the final product, the mature, or relatively fully-developed mind, be understood. Anatomical researches were long conducted on the bodies of animals before the light thrown on structure by embryology cleared up the obscurities which of necessity hung about parts, the origin and early development of which were unknown.

Comparative anatomy had already done something to give increased significance to anatomy as a whole, but it was only by tracing the animal body back to its primitive germ cells, following these cells in their development into tissues and organs by the naked eye and with the microscope, comparing these changes in one animal with corresponding ones in another, and indeed in plants, and interpreting them all in the light of evolution, that the present status of biology has been reached.

Psychology is as yet in no such position; but it

must be equally clear to those who, guided by facts alone, untrammelled by tradition and dogma of every kind, compare the pyschic status of the young with that of the mature animal, that psychogenesis is a fact; that the mind does unfold, evolve, develop equally with the body. And as with the body so with the mind, each stage in this development can only be understood in the light of all the previous stages.

This truth is apparently as yet only dimly comprehended, for, till recently, studies on psychic history, development or psychogenesis have been all but unknown, and as yet, even in the case of man, are very few and confessedly imperfect.

But just as we have an ontogeny and phylogeny, just as the anatomy, physiology, and pathology of man are clearer from comparative studies on creatures lower in the scale, so must it be in regard to man's psychology.

It follows, then, that all researches in comparative psychology must be as welcome for the general science of mind, and the special study of human psychology, as those in comparative anatomy are to anatomy in general, or the anatomy of man in particular.

Till very recently animals below man seem to have been almost wholly neglected or misunderstood in all that pertains to their psychic nature, one very obvious result of which has been the inability to connect the psychic states of man with others of similar, yet often simpler, character in lower animals, not to mention the impossibility of a science of mind in general, or a true understanding of the psychic side of man's nature. Studies in infant psychology are of comparatively recent date, few in number, and in most instances very incomplete; while, as regards animals lower in the scale, such investigations are still more imperfect.

The relations of mind and body in both health and

disease have been made the subject of considerable speculation and some valuable research, but the subject is vast, and will unfold but slowly till our knowledge of many things is greatly increased.

Much depends on the philosophical or scientific attitude of the worker, as to the views he holds on such a subject, or the interpretations he puts on observed facts.

Nevertheless, to him who can lay aside prejudices—sanctioned, it may be, by ages of belief—it is possible to see that old interpretations fail, and that problems of the mind, which the world has either ignored or grappled with in vain, must be attacked from new standpoints.

History and Objects of the Present Research.

In consequence of the foregoing and many other convictions, some ten years since, I suggested to the students of the Faculty of Comparative Medicine and Veterinary Science of M'Gill University the desirability of forming a Society for the study of Comparative Psychology, more especially for the study of the psychic nature of those animals with which they would be professionally most brought into contact. During this period, more than formerly, I myself bred and reared large numbers of the smaller of the domestic animals and pets with a view of understanding them in all their varied aspects.

The longer, however, I continued my studies, the more I became convinced that, as in every other case to succeed best, one must begin at the beginning. Accordingly I have for a few years kept full, and I hope accurate, notes of the development, psychic and physical, of individuals belonging to several different groups of the above-mentioned animals.

My purpose may be stated about as follows:—

(1) To give a detailed history of the psychic development up to a certain age of representatives of several animal groups.

(2) To compare groups and individuals.

(3) To correlate the psychical and physical—or, at all events, to make some attempt to connect, in time, the psychic and physical development.

The completion of this work will even, so far as I am able to accomplish it, take a considerable time yet, so that I shall be obliged, in the present paper, to confine myself to one group of animals, viz. dogs, of which I have made a study during the greater part of my life, and more especially within the past ten years, as regards their psychic nature and certain other features.

The present paper will be founded chiefly on the notes or diary of three litters of puppies—two of the St Bernard and one of the Bedlington terrier breed.

These histories, then, will concern, it will be observed, only pure-bred dogs, as I have not as yet similar notes on mongrels. As the dog is, after the monkey, more like man psychically than any other animal, I hope to make some comparisons with the development of the young human being, though possibly not in this paper.

Inasmuch as the diary of the last litter of St Bernard puppies studied is more complete, and was written in the light of my past experience, I regard it as much the most valuable. It will therefore be given first of all, as written day by day, with only a few verbal alterations, from which each reader may form his own independent conclusions.

This I purpose to follow by certain remarks. As my work on the brain especially is not yet complete, the physical correlation which has to do chiefly, of course with the nervous system, will be less fully treated than the psychical development.

Diary.

The following record concerns a litter of pure-bred St Bernard puppies, whelped in my kennel in the spring of 1894. Both sire and dam were of excellent breeding, and the pedigree, for many generations, was known. The dam had a gestation period of about sixty-one days, so that the puppies may be considered to have been born at full time, and they were certainly very strong and active. They were of unusually even size, and with little apparent difference as to vigour, etc. There were seven of the male and six of the female sex, all of which were not preserved; for some time, however, there were nine, and to the end of the sixth week seven; after that six.

The dam whelped in a separate compartment of the kennel where she was all alone and free from disturbance. The arrangement to meet the comfort of the dam and her offspring, which I will term the pen, was as follows : On a floor, slightly raised above that of the kennel, some clean, dry straw was littered, the whole being surrounded by a board enclosure to the height of about 1 foot. This pen measured about 3 by 3 feet. Care was taken to change the straw on the floor, while the whole kennel was well lighted, comfortably warmed, and properly aired. The dam was given the best of care in all respects, never had an unfavourable symptom during or after whelping, and was always able to furnish her offspring with abundance of good milk. For many reasons these details are of importance, and it is necessary to state them in order that the record may be properly appreciated. Nearly all the observations for some weeks were made on the puppies in their birthplace, as it was found that removal therefrom caused so much disturbance that observations were impossible or valueless except to illustrate this very point, important in itself.

I have limited this diary to the first sixty days of life, as nearly all the most important phases of development show themselves within this period.

1st day.—Almost as soon as born and freed from the investing placental parts by the dam, the puppies *cry* out, though more loudly a little later, *crawl* slowly but vigorously enough towards the teats of the dam, and at once, in most cases, begin to *suck*. It is noticed, however, that other parts are sometimes sucked as well as the teats. They huddle together and get between the legs of the dam, and where the hair is longest, or where, for any reason, there is most warmth, when not actually nursing.

Their movements are very slow. Their eyelids are still not grown apart nor their ears grown open.

Two of them weighed at the end of about twenty-four hours 1 lb. 2 oz. and 1 lb. 6 oz. respectively.

They were not examined as to reflexes other than sucking, reaction to temperature, etc.

I made, on the first day, the following experiment: Placing a puppy on a surface above the floor, it was found that, when it reached the edge, it became very *uneasy*, spread its claws, grasped, etc., to avoid falling off.

On this and later days they cry apparently from cold or hunger, or when removed from the usual environment.

4th day.—The last experiment is repeated under slightly varying conditions. A tortoise placed under the same conditions walked or tumbled off. On this day one puppy was conveyed to my laboratory, wrapped up warmly in a blanket, without a cry or other sign of discomfort, this journey occupying about half an hour.

5th day.—When pinched, they gave evidence of *feeling* by a cry and movement, though the latter is not very marked. When the hand is laid over them in a caressing way just afterwards they are at once quieted. I regret that this experiment was not made earlier.

6th day—Several attempts are made to ascertain if they smell, but with uncertain results. Warm milk and meat were held near their noses. I think there was some sniffing as a result, but cannot be certain.

7th day.—Growing well. Two specimens (females) weigh 2 lbs. 7 oz. and 2 lbs. 10 oz. respectively. Tested taste by the use of milk and of aloes. A finger dipped in milk is long sucked. When aloes, in solution, is placed on the finger, the latter is not long sucked, and the facial movements indicate *disgust*, quite the reverse in the case of anything sweet. I endeavoured to learn whether they knew if the dam were near them by smell, but could not establish it. Up to this date, and long after, no evidence of hearing to be elicited.

9th day.—When the dam is out of the pen, as she now often is, the puppies suck frequently at different parts of the bodies of each other. They will suck vigorously and for some time at my finger.

It is easy to notice now great progress in power of *movement*, especially as regards the forelimb, mouth parts, and head or neck. No movement of the tail at all yet nor for some time.

10th day.—I again attempted to determine whether they could smell, in the same manner as before, but with no definite results, though strongly inclined to believe that they could to some extent.

When the dam, after an absence, steps into the pen, two or three may happen to get between her legs after she lies down. Presently these and others commence to move in a lively way in all directions, and before long manage to reach the teats.

11th day.—Held a saucer containing warm milk under the nose of one of the puppies. It took the edge of the saucer in its mouth. Another tried to drink the milk but did not succeed, its eagerness being in excess of its ability to co-ordinate muscular movements.

The evidence of smell is still very doubtful.

It is now easy to discern that some are larger and in better physical condition than others.

13th day.—Last night it was observed that the eyes began to *open.* At noon to-day they are not fully open, being held by a thin sheet of tissue at the outer canthus; individual differences are very marked, however, in this matter.

Smell is tested with pieces of cold cooked veal, warm fried kidney, and cold cooked salted herring. All, when these were put near the nose, *licked* their lips and moved forward and to each side, following the objects evidently by the *nose*.

They do not wink when the whole hand or a finger is moved close before their eyes, but when the eyelashes are touched, or all but touched, they do wink. The same reflex follows where the lid or corner is actually touched.

It is very difficult to make out the pupil, and I was not able to learn, though I tried day after day, whether the iris contracted to light or not. No evidence of the existence of vision could be obtained.

A feather inserted into the nostril causes the head to be quickly drawn away.

Considerable twitching of the muscles is noticed when they are asleep.

There is a tendency to *growling* in sleep.

All the movements are better than on the earlier days; and for the first time slight *tail movements* are noticed, none having been observed during the period prior to opening of the eyes.

When a puppy is removed from the others in its pen it manifests little *uneasiness*, but quite the reverse if placed on the floor of the kennel, which is covered with sawdust. It creeps about and cries.

14*th day*.—Unable to get any evidence of seeing objects, as no sign is given of any kind when various things are moved before the eyes, nor is the winking reflex any better established.

They seem, as before, to crawl against the board wall of the pen without noticing it. The eyes are more fully opened.

The loudest noises, including the sounding of a shrill dog-whistle, that can be easily heard a quarter of a mile away, causes no reflex movements of the ears, or any other movement to indicate the possession of hearing. On the other hand, a slight breath of air causes reflex movements.

To-day I made a definite test of the *temperature sense*. A glass pestle was heated till it could not be comfortably borne on my skin anywhere, when its end, about half an inch in diameter, was placed against the paw of the puppy, which was rapidly withdrawn. A similar reaction followed the application of ice, but not so quickly.

They now begin to use the *jaws* apart from sucking. They stand better and move faster, the hind limbs being, however, much less under control than the *front legs*.

I suspected that the beginning of play appeared to-day, but was not quite certain.

The tendency to *growl* is manifesting itself in sleep.

15*th day*.—Eyelids continue to grow apart, so that more of the globe of the eye can be seen. They seem to wink reflexly a shade more readily under the former tests, but more promptly with the finger close to the eye than with the entire hand moved as close as possible before the face.

One puppy appears to see, or be trying to see, the dam, judging by the position of the head, etc., but it is possible that it is partly guided by smell. I allowed the dam to stand just within the pen at some little distance (1 or 2 feet) from the puppies lying asleep or drowsy. An uneasiness was manifested which increased, and was probably due to their smelling the dam.

On bringing some sulphuric ether towards the nostrils of one of the puppies there was decided evidence of dislike.

When they are lying asleep, touching the lips gently causes movements of the muscles of the face, and especially of the tongue—an incipient sucking, in fact. All tests of hearing give negative results. It is impossible to introduce a small probe into the auditory canal—which attempt was made with the puppy under ether so as to avoid the shaking of the head, which might introduce fallacies, and be a source of danger to the drum-head of the ear.

For this and other investigations that could not be well carried out at home, one of the puppies was conveyed to the Physiological Laboratory of M'Gill University. The puppy, having sucked to its satisfaction, was tucked up warmly in a basket, and conveyed for twenty minutes in a street car without the slightest signs of uneasiness.

Whenever the puppy recovered the least from the ether anæsthesia it showed a tendency to whine, cry out, move, etc.

To-day there was undoubted *play* witnessed, both paws and jaws being used, especially the latter. The second subject participated to a less degree. There was no sucking of the ear or other part of the body in this case, as had often happened before, when the mouth of one canine casually came in contact with the ear, paw, etc., of another puppy.

16*th day.*—Can discover little advance in vision. The eyes are still more fully open.

If the puppies hear at all it is only in the faintest way.

As the dam stands close beside the pen, when the puppies lie drowsing, they soon begin to move the muscles of the face, raise their heads, sniff here and there, like a hunting dog catching scent of game, and feel about, as it were, for the object giving the scent. Presently they make, almost simultaneously, quick movements as if to reach some object. I am convinced that all this is from smell and not vision, though it would be difficult to prove absolutely that sight had nothing to do with it.

When the dam now sits on her haunches the puppies manage to reach the teats.

They will still suck the finger when put into the mouth, but for a much shorter period.

17*th day.*—Playing more common. One began to play with the foot of another, but soon changed to sucking this part.

Slight movements of *the tail* are noticed during play at times, and there is obvious increase in walking power; muscular co-ordinations of all kinds are better made.

When a beef bone is held within half an inch of the nose when the puppies are asleep, the movements of tongue, lips, etc., before referred to as evidence of *smell*, take place. When awake, they give evidence of smelling cold roast beef at 3 inches.

When the dam stands at the end of the pen, some 2 feet from the puppies, that lie in about its centre, they soon begin to move towards her, but not in a straight line, as they would if they were guided solely by sight. I am convinced that vision is very imperfect yet.

The *ear-flaps* have, for a couple of days, been turned forward instead of backward as at birth, but tests for hearing give but uncertain indications as yet.

By touching, in a certain way, either the outside of the flap of the ear, or its inside and adjacent parts, the *scratching reflex* is excited on the same side. Occasionally the puppy attempts to get rid of the irritation by the use of the foreleg of the same side.

Upon suddenly seizing one of them it *growls*.

The *winking reflex* is more readily obtained, and the latent period is shorter.

While the puppies may have some vague notion of the existence of objects by their eyes, no clear evidence of being able to see objects in the proper sense of the term is to be obtained, notwithstanding many attempts.

For the first time they *lick* the finger without any attempt to suck when it is presented to them.

They *swallow*, but not very well, a little fluid placed in the mouth, though they retch when the handle of a spoon is placed far back in the pharynx; this is neither very pronounced nor very sudden. Upon putting the finger to the front of the mouth, the foreleg is used to remove it without any attempt at sucking. (Will or reflex?)

Up to this date *exhaustion*, under any stimulus, is a marked feature to which reference will be made subsequently.

It is again noticed that all reflexes are more perfectly carried out, and the latent period shortening.

When the dam was nursing the puppies one of them was put behind her. It felt about for a short time, and then got round to the front and soon reached a teat. Another did the same, though not so quickly.

Individual differences are now more evident.

Certain important points were settled on the evening

of the 17th day, to which special attention is called. The observations and experiments were made at 8.30 P.M.

During sleep, growling, twitching of the muscles, sucking movements, licking, etc., are observed.

The dam was placed near the puppies when asleep. There was licking of the lips, and general uneasiness, but no actual waking up till the dam stood near the centre of the pen where they were lying, when some stood up, and were evidently "feeling about for the body scent," as sportsmen describe the action of their dogs when they detect the scent left by the bodies of game birds as opposed to that of the feet.

Special tests were made of hearing. Clapping of the hands rouses them suddenly, but not in that unquestionable way seen later, for wafting the hand over them does the same, but less suddenly.

Then low growling and low barking sounds are made, which seem to rouse them at first a little, but this was not demonstrative.

Upon sounding the dog-whistle loudly there was a doubtful twitching of the ears, etc., but on repeating any of these tests the results were still more doubtful or wholly negative.

To determine whether this was due to the concussion caused by clapping the hands, or to the actual aerial vibrations, the physical stimulus in hearing, I stamped on the floor where I stood when clapping the hands, causing more concussion than the clapping possibly could, but with no results.

Finally, a thick cloth was interposed between the puppies and the hands, when the result was positive, showing conclusively that *hearing* was now established on the 17th day.

One of the puppies, upon having his back rubbed the wrong way of the hair, or rather both ways, growled.

Several others were tried, but, while roused in a measure, did not growl, and even the first one soon ceased to react.

Although ordinary sounds do not rouse them, feelings of discomfort do, for they rarely or never empty the bowels or bladder now where they lie, but move aside to do so. As in the case of rubbing the back, reflex effects get fainter and soon cease.

18th day.—Being without food for three hours the puppies are very active. They walk about with tails up, and play with each other.

In order to determine whether they are still guided by the sense of smell or by sight also, two of the puppies were removed from the pen and their eyes bandaged, but this seemed to confuse them and render them so uneasy that no conclusions could be drawn.

However, when they are held up before a good light, they follow with their eyes the movements of the hand or other object; nevertheless, when they reach the dam from the distant part of the pen, it is difficult to determine how much they are guided by sight and how much by smell. I am convinced that, while the former is an aid, smell is still the most useful to them in all such cases.

The peculiar noise made with the lips to attract the attention of dogs, which I may term the *lip-call*, is evidently heard, and as the position is shifted the puppies follow the sound to right and left. While the dog-whistle is heard, it causes reflexes of the ears and some startling, but does not rouse them so thoroughly into movements as the lip-call and certain other sounds.

When an attempt is made to plug the nostrils with cotton wool, it is at once sneezed out reflexly.

Judging by the whining and crying after fasting, hunger is more keenly felt than ever.

A bandage placed over the eyes soon causes sleep.

Up to the present date the only nourishment received has been the mother's milk, but to-day artificial feeding with cow's milk diluted was added. The first attempts at lapping, though far from perfect, were fairly good— much better than the first attempts at swallowing fluid artificially introduced. It is noticed that they follow up slowly the spots where milk had been spilled. After each feeding they lick each other's faces thoroughly.*

19th day.—The attendant reports the puppies as *barking* when he entered, as if at him.

The lip-call, at a distance of 6 to 8 feet, causes them to prick up the ears quickly, which is soon followed by crying (expectancy of food possibly).

One of the puppies *scratches* his own ear.

Teeth are appearing that for some days could be felt beneath the gums.

20th day.—Some get additional teeth.

21st day.—Certain motor manifestations are worthy of special mention.

Tails are wagged during play, and walking with tail held *erect* is seen for the first time.

Several of them tried to get out of the pen.

When the muzzle is held by the hand both hind legs were used in an attempt to remove it. (Will or reflex?)

The hand moved before the face, as if to strike, causes winking.

Now they seem to *hear almost every sound* made in their compartment of the kennel, which is about 15 by 6 feet.

22nd day.—Some have all the upper incisors, and in

---

* During four days I was absent from home, but the puppies were carefully watched and notes taken by members of my family, who are familiar with the ways of dogs, and had frequently been with me when making my investigations on this and other litters of puppies.

one case the nose is all but covered with the characteristic black pigment, though this one is in advance of the others in this respect.

23rd day.—On my return on this day a long time was spent with the puppies, and the following noted:

The dam is no longer so much inclined to stay with her offspring, and does not wish to suckle them so frequently.

The puppies are fed on the top of a large box, two at a time. It is found that unless straw is placed on the top of the box the puppies will not feed. They decline to take half milk and water any longer, but must have richer food, and considerable attention must be paid to the temperature of the liquid.

Great improvement is noticed in lapping milk, though one is observed attempting to gulp the milk as it were (hunger, etc.).

Soon after feeding, the finger placed in the mouth is not sucked but rather *chewed.*

The readiness with which all sorts of sounds are heard, even when some distance away, is striking.

The puppies now follow a small object or a piece of paper (2 by 3 inches) held within a few inches of the face.

Much *growling* in play, also more advanced use of *tail.* They also wag the tail now sometimes when an object is presented to them, or when anything pleases them. They turn the head quickly towards any part of the body gently pinched.

On pinching one of them frequently and rapidly much *irritation* is shown by the voice, expression of face, etc.

They now very frequently stand with the paws on the edge of the enclosing boards of the pen, and show that they would like to get out. The height of the pen is now about 15 inches.

I notice one sleeping and another licking its face after feeding, using its paw with *movements* closely akin to those of the fore-limb against the mammary glands when sucking.

I can observe a very considerable advance in the use of the *hind limbs* in walking in four days.

During the night one of the puppies had got out of the pen and was making loud outcry and trying to get back.

24th day.—Special tests as to sucking finger gave the following results:

Some on one occasion suck the finger, others do not.

Later, three were tried, one asleep or almost so, the others not, but all sucked the finger tip.

One lying sucks the finger, and puts up its fore feet towards the hand, and *spreads the claws*, at the same time moving the hind limbs somewhat.

One, when standing and sucking at the finger, also lifts its paw.

25th day.—A piece of meat held before the nose of a sleeping puppy at a distance of $2\frac{1}{2}$ inches wakes it (smell). When this piece of meat is rapidly moved before the face at 3 inches, it is as rapidly followed by movements of the head. Was this owing to smell, or sight, or both?

When the meat is put into the mouth it is not merely sucked, but an attempt is made to *chew* it.

When the hands are clapped sharply once, starting is produced, suggestive of more than a mere reflex—possibly real *fright*. When I whistle somewhat lightly some of them *bark*.

26th day.—When I whistle at the distant end of the pen they bark, some of them, but, employing the lip-call, they move in that direction.

Moving a small piece of rag before them as was done with the meat yesterday, causes similar corresponding rapid movements of the head, and it almost seems as if they have some of that sense of fun, or whatever it may be, that we witness in older dogs under similar circumstances.

They can now follow a small object at the distance of at least 1 foot; and at 5 feet they can follow the movements of an object the size of a table-napkin. Both the eyes and head are moved.

On striking a single blow on the bottom of a watering-can, they all rushed off to the distant part of the pen, with all the expressions of fear.

They are now well supplied with *teeth* in both jaws, but in regard to this also there are individual differences.

They *play* much more.

Being rather chilly to-day they huddle together.

Same day at 7 P.M.—When all are playing a slight but well-defined sound causes them all to stop at once.

When the hand is put down in front of them, after they recover, one comes up wagging the tail.

The *eyes* are now very widely open, the *expression* changed, and they can follow the movements of a table-napkin at a distance of 6 to 7 feet, but winking by the old test is not appreciably more pronounced.

One of the puppies, when placed on the floor of the kennel covered with sawdust, plays about, cries, and is evidently very uneasy, if not confused. When put on the top of the box on which they are usually fed, it sniffed and looked towards the white plate from which they drink their milk. When held in the arms it soon manifests uneasiness; when placed near the edge of the box it grows very uneasy, but does not jump off. Almost at once, when placed back in the pen, it became quiet, and soon began to play. By its movements it indicates clearly that the *direction* of sound is perceived.

27*th day*.—The puppies follow a small object (3 by 1 inches) dangled before them at 15 inches.

One is observed playing with a straw 3 or 4 inches from it. In this act there is the use of the mouth and the fore-limb, with all that this implies.

There is clear evidence that sounds made at the outer door of the kennel, and in the adjacent yard, are heard.

A basket in which meat had been kept, and giving off a strong odour when brought near the pen, is plainly smelled. The dam is brought within 3 feet of the puppies, but unseen by them. First one, and then another begin to sniff and soon to cry.

The playing shows advance; better use is made of the hind limbs, which develop functionally and much more slowly than the forelegs. The head and neck movements are also better in all respects.

Now, and even some days since, increase in the quantity and quality of the *coat*, with changes in the shape of the *head* are evident; and in both physical and psychic characteristics, *individuality* is to be noted.

To-day play seemed in one case to change into a little *quarrel* for a few seconds.

One is observed to utter an abortive *bark* in its sleep.

28*th day*.—Noises above the kennel in my pigeon loft have greater effect on the puppies than on the mature dogs in the adjoining kennel.

They can now follow with the eyes the small objects used in all these experiments at a distance of 4 or 5 feet; while a napkin, etc., can be followed anywhere within their kennel compartment.

Various objects, as a plate, glass, a folded napkin, and a Spratt's dog biscuit, are presented, but they mouth all about equally, so that distinct selective choice is not shown.

A small Bedlington terrier bitch that had never had puppies was placed amongst them. All rushed around her and *tried to suck* her undeveloped teats. Then a St Bernard bitch, nearly as large as their dam, was placed in their compartment. Though from their mode of sniffing it appeared that they recognised this animal as a stranger, they soon tried to suck her also.

When they are spoken to in a friendly way they wag the *tail* and give other evidences of sociability by the *face*. They get up on the edge of the pen with forelegs when either the dam or any person is about to leave them, and follow with the eyes, and evidently would with their limbs if they could get out.

They are not now nearly so easily fatigued by any stimuli, being able to last out three or four times as long as they could eight days ago.

29th day.—Puppies follow a small object at a distance of 7 feet.

A *high-pitched, peculiar* sound causes ear reflexes and barking, while a low-pitched sound, imitating barking, has very little effect.

When a small piece of cloth is dangled before the face of the puppy, it tries to catch it with the mouth, and raises one foreleg at the same time, as if to assist in this.

Noticed well-executed *scratching*.

They seem thus far to prefer milk to broth or meat.

30th day.—I did not make special notes of observations on this day.

31st day.—It seemed that to-day the dam was undoubtedly *recognised by sight* alone.

When a bone and the napkin used in the last test of this kind were presented to the puppies, each one at once *selected* the bone. No chewing of the napkin, which shows a distinct advance since the 28th day.

They now observe a small object at any part of their kennel compartment, *i.e.* at 12 to 15 feet.

One was noticed watching with an intelligent expression the movements made in connection with photographic apparatus within 5 feet of them.

When the lip-call is uttered they wag the tail like older dogs.

32nd day.—By lamplight a puppy follows by his eyes a *straw* moved before him at 3 to 4 inches distance. He also seems much interested in the *shadow* of my hand on the wall.

On holding the coal-oil lamp near them, all licked at the glass cistern containing oil (smell). One or two touched the chimney with the nose or tongue, but the majority turned away when it was near the nose, while neither of those that had touched the chimney went near it again.

They show *sociability* with human beings, and a tendency to play with them.

They become very quiet and *attentive* when they hear certain kinds of sounds, which is prolonged if the sound continues.

33rd day.—When a straw is rapidly moved before them they snap at it to catch it.

Upon changing the straw bedding in their pen they *rub about* in it much as old dogs, evidently well pleased.

There is a very distinct advance in the ability to *lap* milk.

Now, when put on the covered floor of the kennel, with its covering of sawdust, they do not manifest uneasiness as before, but walk about and play. One is seen to *run* at a slow rate, with his tail up, and several make quick starts forward and backward. On giving the lip-call, and snapping my fingers, one, a few feet distant, ran towards me.

They now *watch* what is being done near them somewhat attentively.

A slight tap on some boards above them causes them

all to move quickly away with tails down and other expressions of *fear*.

34th day.—Tested them with a napkin and Spratt's dog biscuit as on the 28th day. Now they all show very decided *preference* for the biscuit, which is not so attractive as a bone to any dog. One or two began to *smell* about the floor of the pen as an old dog does, and yesterday one was observed *scratching* at a spot on the floor where some excrement had been.

At this age puppies have very sharp teeth, and it is not very uncommon now to hear one cry out when his fellow uses his jaws too freely in play.

Scratching is more common.

While the *winking reflex*, from a simple movement of the hand as before, is not readily produced, a sudden tap on the ledge of the pen will cause winking if they are looking that way, and within a foot or so of the spot struck. They also wink when the muzzle is suddenly touched.

35th day.—They now *retire* to one end of their pen to answer nature's calls.

They are noticed smelling at the shoes of any one who happens to be near them.

They *bark* in sleep like older dogs.

When the finger is thrust into the mouth some suck a good while, some not at all.

36th day.—They follow me around their kennel compartment, and are inclined to *seize* the skirts of a very long coat I wear. By way of testing recognition of the dam, she and her other puppy, eight months old, and nearly as large as herself, were brought to the puppies together. They all at once *rushed* to the dam. But soon after her removal they attempted to suck the younger bitch, though from their sniffing it seemed to me they noticed her strange.

Later in the day the small bitch (Bedlington terrier)

used for a similar test before, was placed among them. They soon tried to suck her teats, at which, on account of her smaller size, they could readily get.

37th day.—Being a warm day the puppies feel the heat a good deal, lie far apart from each other, and pant with tongues lolling out.

At a distance of 10 feet, a mere word uttered in a low voice *rouses* one that is drowsing.

Out of five tested only one sucked when the finger was introduced into the mouth.

38th day.—They gave evidence of seeing me well, though I was standing at an outside door of the kennel, with two wire-mesh partitions between, and at a distance of about 12 feet.

Upon dangling a rope over their heads one seizes and pulls at it, but when doing the same with a bright chain they make off, showing fear. This was probably owing to the noise it made, the brightness, and, in one case, to the puppy having come in contact with it.

The compartment in which they are kept is closed by a heavy wire-mesh door, through which every exit must be made. They *crowd around* this often now, and sometimes whine there when hungry.

39th day.—High temperature; puppies very uneasy.

Seeing me at some little distance, one of them *wags the tail* like an old dog, showing its sociable and friendly nature.

When one speaks they show pleasure by the tail, expression of face, etc.

Two bones from cooked meat were placed on the straw of their pen, which now has walls only a few inches high, so that they can go in and out easily.

One or two go towards the bones, followed by others; one seizes a bone and walks out of the pen with tail up, much in the manner of an *older dog*. I suddenly removed the bone, when the puppy that had it sniffed about, going back over his track, evidently guided by the scent it had left.

Some of them that had followed up the puppies that had taken the bones return to the straw, seeking them.

The removal of a large piece of tin that has been used to encircle a stove in a protective way, made a noise which caused them all to rush away as so many sheep, but when I gave the lip-call they soon recovered and came towards me.

Given *water* in a vessel to drink for the first time, they merely dip into it.

40*th day*.—Being decidedly cooler they do not whine or cry, but play much.

One is seen violently shaking a piece of paper that was in the straw.

Another is seen *scratching* his head, with the latter inclined towards his leg in the manner of a mature dog, though with much slower movements.

To-day one is seen to *lap water* when it is poured into the vessel attached to the wire-mesh partition, whereupon several *others* do so.

In the evening one is noticed moving about in a way peculiar to *an old* dog prior to defecation.

When any one enters the kennel the puppies now *run about* his legs eagerly.

They have almost deserted their pen, and lie about on the floor of their kennel compartment, finding it cooler, while the layer of sawdust makes it soft to rest upon.

41*st day*.—Their pen was wholly removed to-day, as it served no good purpose.

They occasionally lie so that the head and body is in a fashion supported, *e.g.* against the partition or walls of the kennel, or with the head on a part of the floor that is there a little raised.

A very slight growl at the outer door of the main

kennel with three partitions or walls between (doors being open) causes one of them, though lying apparently asleep, to get up, and if anything happens they all awake if one moves much or whines.

One of them is observed to *snap* at a fly.

At 10 P.M. I notice one of the puppies *scraping* away the sawdust near the elevation referred to above, on which he had laid his head in preparing to rest. He tried the spot once or twice before he finally laid his head down.

42nd day.—It is very warm, and the puppies whine and cry a good deal owing to the discomfort, as their coats are thick and warm.

Now there are many evidences that they *hear* as acutely as mature dogs, if not more so, and sounds disturb them more, as they do not know their meaning so well.

They now show an *interest* in everything that goes on within their field of examination with eyes, nose, etc. In fact it is difficult to move about among them.

When they see one they may cry out if hungry, wag the tail if recently fed and satisfied, sniff, etc.

This *sniffing* is a characteristic method of investigation with dogs, and its appearance, at this date and earlier, is significant.

On every occasion, if they see or smell the dam (that is seldom with them now, as they were gradually weaned—the process ending to-day), they cry out.

While this litter is an unusually even one in physical characteristics, at all events, *individual* differences are to be observed in many directions. There are some decided differences in psychic manifestations.

One, a bitch, seems to be quicker and more precocious than the rest by a great deal.

One dog *growls* when feeding, as they do at present all together from one large dish.

43rd day.—It is warm, and two are noticed lying in a *darker* and more secluded part of the kennel, where there are fewer flies perhaps.

It is noticed that now one often *acts as does another;* one seems to take its cue from another.

44th day.—So very warm, the puppies are prostrated by the heat, and lie about, maintaining mere existence.

45th day.—To-day, for the first time, the door of their compartment was left open, so that they might enter an adjoining one, which is in general fitted up in the same way, so that the environment is substantially the same.

It was curious to note the results. It was some minutes before the puppies, the precocious bitch excepted, realised that the door was really open, and that they had free access to a new compartment. They did not at once surmount the difficulty presented by the door case only a few inches high. When some of them came in and saw the water *vessel* attached to the other side of the wire-mesh partition, they did not at once comprehend that they could not drink from it, when they saw their fellow on the distant side lapping. All this, however, lasted but a very few minutes. Soon they all were busy *investigating* the new place with nose, eyes, feet, etc.

The new experiences evidently afford them unusual pleasure in spite of the heat, as they play more than for some days.

To-day I first used a switch to learn what effect it would have on their crying, etc.

They seem to make the mental *association* to some extent, but only imperfectly.

One of them, as they crowded around me, was trodden upon, and this had a decided and somewhat *lasting* psychic effect, as will be seen later. As he was running away after this accident, I caught him, and was trying to soothe the creature, but this was not at first understood, and increased its terror.

Later in the day they understand the whip better. I notice what may be termed *wanton barking* as well as that which denotes but an excess of good feeling—"animal spirits."

Now and then one turns round on another that is attempting to play with it in quite a *fierce* way.

They are running more than formerly.

There are physical changes and good growth notwithstanding the long succession of hot—to them very hot—days.

One of them is observed using both paws to scrape away the sawdust from a part of the kennel floor. He then puts down his head and tries the spot. This was repeated three times before the puppy lay quietly at rest.

Actions of one are followed by *similar actions* in others much more frequently and readily than before.

46*th day*.—It has been raining; the sky is dull and the atmosphere is moist, and though it is still warm the puppies seem less restless and uncomfortable. They cry much less.

47*th day*.—I notice that the precocious bitch acts towards the whip much as an *old dog* or a half grown one often does. This is difficult to describe. The animal shows that it understands what its relations are, but seems to combine a sort of pleading with humour. It is complex, however, and must be witnessed to be understood.

The individual that was trodden upon now *retires* to another part of the compartment when I appear; there is evidently a very unpleasant association of ideas.

At 11.30 P.M. I went to the kennel to see how all my dogs were, as the night was very close. The door directly opposite the puppy-kennel was open. I had no light and walked softly, yet two of the puppies, lying against the wire-mesh partition, some 6 or 8 feet from where I stood in the darkness, awoke and soon began to cry as I passed close to the closed main kennel door. The old dogs were heard sniffing. They evidently detected me by the sense of smell.

Was it wholly so in the case of the puppies, or were they assisted by sight? Hearing may, I think, be excluded, though not with perfect confidence, so sharp now are their ears.

In any case, this observation is of much significance, even be it granted that they were not asleep at the time I stood before the door of their kennel. It is further to be remembered, in this instance, that by a misunderstanding the puppies had not had their last evening meal, and also lacked water.

48th day.—It remains warm. The flies are troublesome, and as the puppies lie asleep, or trying to sleep, the same *movements* of the skin of the head, of the ears, etc., may be seen as in mature dogs when flies irritate these parts.

49th day.—To-day, for the first time, the puppies were removed for a time to a part of the yard enclosed by wire-mesh. The earth furnishes a fresh surface with various small objects on it.

The puppies proceed to *investigate*, as when before they were given free access to new surroundings.

They seize and carry small objects, which they take from each other, indulge in play, and evidently experience keen enjoyment. After, say, half an hour, they lie down and sleep.

When I call "Puppies!" from a verandah, at a height of about 20 feet, and at about the same distance on their plane, they look up, some of them at least, at once, to my surprise, for I expected they would not be able to detect the direction of the sound so quickly.

The bitch puppy was taken upstairs in my house to-day to be weighed. She, like two of the dogs, seemed *abashed* by the new surroundings, but soon recovered, and when some one entered by the front door downstairs, one storey, turned the head in the *direction* of the sound.

50th day.—When crying this morning, one of them was well whipped, with the result that it remained quiet for some hours after. Dogs, young and old, easily acquire habits, good and bad, and barking and crying are examples, and sometimes one or two whippings that are felt puts an end to what renders the dog wretched, as well as those who must listen to him, hence the treatment alluded to above.

In attempting to give them some bromide of potassium to quiet their uneasiness, it is found that they *fight against* the unpleasant stuff, and it is with difficulty they can be made to swallow it at all.

51st day.—They are awake very early (4 A.M.), and eager for food and exercise.

I moved a whip over one that had been making a good deal of outcry. She looked as if she knew what it meant (had been before whipped two or three times). As I moved the whip she put up one *paw* as if to ward it off.

52nd day.—Cooler to-day and the puppies are quieter. *Barking* now frequent; seems to be partly from excess of animal spirits, and at other times from a sort of wantonness. I notice an advance in *co-ordination* in scratching; they adapt one part to another still more like an old dog than formerly.

53rd day.—When I lift the whip, and wave it 3 feet above them, another lifts a paw. They all look as if they knew the meaning of a whip better.

When I shouted "Puppy!" from an upper verandah, about 50 feet distant, two of them that were lying quietly in the sawdust arose, and looked towards the source of the sound.

54th day.—The same sort of pawing, and at the same places as before.

It is scarcely possible to go into the kennel anywhere now when they are awake without some of them *detecting* my presence by ear, eye, or smell, or by all three, as is now evidently often the case.

55th day.—Warm. Much barking and restlessness. They have for some two or three weeks had the range of two compartments of the kennel, but they would evidently like the range of the whole yard as well as the outside run; and if this were once permitted, experience with other puppies has taught me, they might be unwilling to stay in the kennel at all during the day, which condition of things would not in several respects be desirable.

The dog trodden upon still shows that he *remembers*, but will now turn to the lip-call.

56th day.—Eighth week. Though the litter remains an even one, *changes* characteristic of growth and development are evident.

The bitch puppy shows very pronounced changes in colour of coat, expression of face, temperament, etc., and has the most marked individuality of any of them at present. She seems still precocious.

57th day.—They are so active it is difficult to move around among them.

It is noteworthy that they use the kennel compartment they occupied originally as a *retiring place* to answer to nature's calls, while they play, rest, and sleep chiefly in the additional compartment last given them. Perhaps this is to be accounted for in part by the fact that, from the latter, there is a door opening outward, and another of wire-netting, through which they can look out and catch an occasional breeze.

58th day.—An ox's head, that had been boiled free from all flesh, was placed amongst them. They all attacked it eagerly, showing *inexperience.*

An old dog would have acted in this way only in case of extreme hunger.

Some desist and again return to the attack, but show that already *experience* has not been lost on them. Some of them growl when others approach.

59th day.—The puppies are given small rib-bones from cooked lamb. Each carries off his own with tail up; uses the paws to steady the bone; gets hold of it with his teeth by the end, so that he may gnaw off perchance some of it; growls when a fellow approaches, etc. All this was suggestive of the behaviour of an *old dog.* The puppies plainly recognise the nature of a fellow's growl under these circumstances.

60th day.—Temperature higher. The puppies show the effect of the heat both physically and psychically.

61st day.—To-day one sheep's head and a bone for each is placed in their compartment.

In gnawing their own bone, in growling and acting on the defensive generally, there is considerable *advance* over the 59th day.

They are allowed into the large yard to-day for the first time. They have seen this yard from the kennel and from their wire fence run in the middle of it. They mingle with the older dogs and *act* very much like them. They try to suck the dam and both the other bitches referred to before on the 28th, etc., days.

They move about the yard from the first, as if acquainted with it, and choose the comfortably shady places in which to lie. By the lip-call, etc., I get them to follow me back to the kennel, but when inside the door they *hesitate* and soon make for the yard. When placed in their usual compartment in the kennel, after being some hours in the yard, they cry, but not long.

Brief Extracts from the Early Records of the Diary of Another Litter of St Bernards, by the same Dam, but Another Sire.

18*th day.*—First seen playing.

20*th day.*—They seize the finger instead of sucking it. Come at lip-call, with tails up.

22*nd day.*—They no longer mistake other parts for the teats of the dam.

28*th day.*—When called ("Puppy!") they wag the tail.

During the third week (the day not noted) the first attempt at scratching observed.

7*th week.*—Individual differences pronounced.

Brief Extracts from the Diary of a Litter of Bedlington Terriers.

2*nd day.*—Taste tested with Epsom salts and nux vomica. Cannot determine positively whether they either taste or smell.

On the same day *pinching* causes them to cry out with pain, but the *latent period* is notably long.

9*th day.*—*Concussion* of the surface on which they lie causes appearance of fright.

11*th day.*—Eyes begin to *open.*

They smack their lips, etc. (the eyes being covered) when meat is held 2 inches from the nose.

16*th day.*—Ears not well open. Hearing still doubtful. Seem to smell at 3 or 4 inches.

19*th day.*—When asleep I take the dam in quietly. When within 2 feet the puppies begin to move—soon to whine and cry.

Hearing still doubtful, but inclined to think it exists in feeble degree.

This day they managed to get out of the pen, which is 5 or 6 inches high.

They also *co-ordinate* well in scratching.

First *growling* noticed.

Sexual differentiation shown in expression, in shape, and psychic as well as somatic characteristics.

22*nd day.*—Clear evidence of recognising dam by smell when she could not be seen.

23rd day.—*Playing.* Differences in coat, shape of head, etc., showing a physical advance.

They now *bite and chew* at objects. They show a decided aversion to Epsom salts.

25th day.—Began feeding milk. They do fairly well at first attempt to lap.

26th day.—They push through some slats confining them, showing considerable *co-ordinative* power, etc.

30th day.—Repeated and rapid pinching of their sides makes them very *angry*—snarl, etc.

The fall of a shovel causes them all to cower with *fear.*

32nd day.—Lip-call followed by their *approach* with wagging tail.

37th day.—The sound of a whip surprises and seems to puzzle them.

45th day.—When about to punish another mature terrier they hide away under the benches.

They are put down in the yard, a large one, for the first time, and seem *puzzled and shy.*

46th day.—Great changes now visible in physical features, expression of face (more knowing), etc. They now crowd each other when eating from the same dish. Their movements and whole demeanour more terrier-like. This is seen in play very clearly.

They are now much more readily and profoundly affected by noises.

47th day.—Rapid development owing to enlarged *experience;* much more aggressive.

50th day.—Two of them given bones. Each goes off with one. When one comes up to take the other's, he pulls it away but does not growl. Lies down to bone and uses his feet to steady it like a *mature dog.* When the bone is snatched up the puppy *sniffs* about after it.

Sexual and individual differences now more evident—I mean that the peculiarities of shape, expression, and demeanour that characterise a mature bitch, and which only close observers of dogs detect, are now fairly well developed.

58th day.—When looking out into darkness at night they show *hesitation, fear,* etc.

The discussion that follows is based almost entirely on the diary of the litter of St Bernard puppies, extended over sixty days.

The extract from the diary of another litter of St Bernards (their half brothers and sisters) is introduced for comparison chiefly; that of the Bedlington terriers for this reason, and in addition because it supplements the chief diary, and in some respects makes good omissions in investigations in the early days.

Remarks on the Diary, etc.

As the litter of puppies on which these remarks are chiefly based was a very healthy, active, and especially even one, there being no weaklings, and none very much in advance physically or otherwise, the notes are of the more value as representing observations in perfectly normal specimens of pure-bred dogs.

The facts most striking in the first few days of life are the frequent desire to suck, the perfect ability to reach the teats of the dam just after birth, the misery evident under cold or hunger, and the fact that the greater part of existence is spent in the sleeping state. The latter is so well known that I have not thought it necessary to make special notes upon the subject, but it, of course, gradually gives way to a form of existence in which sleep has a less and less prominent share.

There are many reasons why so much time is spent in sleep, and why sleep is so readily induced, to some of which reference has been made in the diary, and to which I shall refer again.

All parts of an animal's body, owing to nervous or simply protoplasmic connections merely, are in relation to each other, and this must constantly be borne in mind if we would understand psychic as well as physical (somatic) phenomena. The nervous centres, however, constitute a sort of head office, or series of offices, where the various changes of the body are reported, correlated, etc., in all higher animals. In the youngest, though the cerebrum is but indifferently active as yet, the lower nervous centres are constantly receiving impulses coming from peripheral parts, the viscera included, and if these are of an abnormal or disturbing character, there result those forms of expression or external representation of the ingoing effects, mostly movements which we can correlate with their causes. Hence the young animal expresses its feelings of discomfort as hunger, cold, etc., by movements, some of which result in cries, whining, etc., and experiments, as well as the behaviour of animals born without the cerebrum, show that the higher parts of the brain may be little concerned.

The feeling of discomfort from being in an atmosphere that is not warm enough, is different somewhat from the sensation, likewise disagreeable, of a body too cold being placed against the skin. Effects not confined to the surface, but modifying the whole of the vital processes, result from the former, as it is well known that very young animals cannot exist at all in a temperature below a certain rather high point as compared with that endurable by mature animals.

Nothing is more striking than the efforts the animal makes almost as soon as it is born to place itself in an environment of comfort. The importance of this instinct—just as fundamental as sucking, etc.—will be evident when one considers that the vital processes cannot continue except under these conditions. It is even more important than that there should be a supply of food within the first few hours.

SUCKING.—Sucking has been so frequently referred to by writers as an example of a perfect instinct, that I have taken pains to give some details regarding it, and to trace its modifications and final decline.

It will be observed by any one who will, without prejudice, examine the subject, that sucking is not perfect at first; that, like the lapping of milk, swallowing, etc., but much less so, it is improved by practice, and that it is subject to modification with the increasing experience of the animal. It is true the mechanism of sucking, both muscular and nervous, in consequence of countless ancestral experiences, is like perfectly made machinery in good order, it will work on the slightest stimulus, but later this machinery is better oiled; it works better. That there is but imperfect discrimination as to what is sucked is well shown by my diary, and that the act only continues a certain time, when milk is not obtained, proves that the instinct is fairly perfect. However, as the notes show, the older the puppy the more perfectly does it utilise the sucking mechanism, the less energy does it waste, *e.g.* the feet are used to much greater advantage in pressing the mammary glands after a couple of weeks than in the first days.

Does the puppy find the teats shortly after its birth by smell? I am convinced that it plays no great part in the matter for some days, as far as dogs are concerned. After birth they crawl towards the mother's abdomen to get warmth; they tend to suck almost any fleshy object that comes in their way that is not cold; they meet the teats, which are the objects best adapted to seize and suck; getting satisfaction, this is continued. No doubt, later, smell, the tactile sense, still later vision, and a whole host of stored experiences guide in this, as in other cases, in which instinct is essential and most prominent in the result. But that smell is essential that a puppy may reach its dam's teats soon after birth I cannot believe, from the many observations I have made.

PAIN.—That a puppy, in the first hour of its existence, feels discomfort cannot be doubted, but I regret that I did not make some definite experiments on the subject of pain on the first day, even in the first hour. This has been made good in part by brief extracts from a diary kept of a litter of Bedlington terriers, previously introduced. Such experiments are necessary, as the discomfort one witnesses in young puppies might be due in certain cases to internal and not to skin sensations.

TACTILE SENSIBILITY.—Very striking, indeed, are the effects on a puppy of any age up to two months (and noticeable even in mature dogs) of stroking, smoothing movements with the hand. In some very young animals, as birds, I find a similar effect, due to placing the hand on them or over them. In this case the effect is largely due to the heat of the hand; in young puppies the gentle tactile stimulus is the principal, but not the sole cause of the quieting effect. In this way a puppy may, when very young, soon be put to sleep, *i.e.* the activity of the nervous centres is inhibited by tactile sensations, so that the frequent lickings of the dam not only cleanse but soothe the puppies. There is, after the eyes are opened, a very rapid increase in the acuteness of tactile sensibility, well shown in the readiness with which a slight touch on the lips will induce motor response, especially well seen in sucking movements, etc.

TEMPERATURE SENSE.—Experiments in this subject were, unfortunately, not made in the early days. However, I tested a kitten, five days old, with an iron warmed and also with ice, getting decisive results of a positive kind. I think that it is likely that the temperature sense is well marked from the first, though the squirming, cries, etc., of young animals are not of themselves conclusive as to this.

THE MUSCULAR SENSE.—On this subject a few words will suffice. Considering how numerous and perfect are the co-ordinated muscular movements of comparatively young puppies, this sense must be early present, and finally well developed.

SENSE OF SUPPORT.—I have found in the case of all puppies, and several other kinds of animals examined, that even on the first day of birth they will not creep off a surface on which they rest, if elevated some little distance above the ground. When they approach the edge they manifest hesitation, grasp with their claws or otherwise attempt to prevent themselves falling, and, it may be, cry out, giving evidence of some profound disturbance in their nervous system.

It would seem that there is no more urgent psychic necessity to young mammals than this sense of being supported. All their ancestral experiences have been associated with *terra firma,* so that it is not very surprising that when *terra firma* seems about to be removed they are so much disturbed. To my own mind this is one of the most instructive and striking psychic manifestations of young animals, though I am not aware that any attention has been called to it before; and instead of referring to it under any of the usual divisions of sense, as the muscular sense, pressure sense, etc., I prefer to treat the subject under the above general heading, for it seems to me that the feeling is a somewhat complex one.

It is interesting to note that a water tortoise I have had for some years will at any time walk off a surface on which he is placed. But this is not a creature that always is on *terra firma* in the same sense as a dog, but it frequently has occasion to drops off logs, etc., into water. But again, I find this sense of support well marked in birds that drop themselves into "thin air." Nevertheless, a consideration of ancestral experiences throws light on most cases, and perhaps on this one also.

TASTE AND SMELL.—These things are so closely

connected anatomically, and especially functionally, that investigations on the one or the other, and particularly on taste, at a very early stage, are attended with great difficulties; accordingly I have been very cautious in drawing conclusions, and have thought it better to place the first beginnings of their exercise too late rather than too early. Certain it is that both taste and smell are very feeble at first and gradually developed. Prior to the opening of the eyes both exist, but in feeble degree. The diary gives all the facts I have to communicate on the subject.

The way in which smell calls into activity, first of all, muscles of the face in a sleeping puppy, has been very frequently brought to my notice, and shows how closely afferent and efferent nervous paths are generally related, even when the main centres concerned are at rather distant parts of the brain. The nervous impulses that pass to the brain when strong enough, soon spread to other parts, hence the puppy is not long in moving its limbs, and, it may be, gets up, runs about, cries, etc.,—all these complicated movements having been brought about, and, as I have often witnessed, in a sort of machine-like way—the animal having no clear and definite features before it at the first moment, though, no doubt, the law of associative nervous and psychic connections complicates this more and more as the animal widens its experiences with age. As illustrating this subject, an observation of mine on a mature dog is worth a brief recital. The subject was an Irish setter bitch of an unusually affectionate nature. I had not seen her for some months. She was lying apparently asleep on her bench in a large dog show. Upon walking up to her stall, and standing there a few seconds, I noticed, the eyes being closed, movements of the nostrils of gradually increasing force, then evident sniffing, next a raising of the head, opening of the eyes, with first of all a dazed sort of expression, then one of great surprise and enquiry, followed shortly by her throwing herself upon me with a bark, almost a shriek of joy. She passed through all the stages the puppy manifests, but with those added ones coming from enlarged experience and a richer psychic life.

The part smell plays in the ordinary and extraordinary life of the dog is a most interesting and by no means exhausted subject, which, though tempting to pursue, is somewhat aside from the scope of the present paper.

As illustrating the development taste undergoes in a few days, special attention is called to the accounts given on the 28th, 31st, and 34th days.

Experiments on taste might have been made at an earlier date, but this omission was supplied in the case of another litter of puppies to which reference will be found in extracts from a diary introduced later.

Some references to smell, as it influences habits, even in very young puppies, have been referred to in the diary.

In the dog, much more than in the man, are smell and taste associated, and this becomes evident in the early as well as the later psychic life of this animal, as shown by the diary, though this is like many other features, much more evident to the one who daily associates with animals, than it can be from the best description it is possible to write.

Vision.—Owing to the gradual opening of the eyes, it is difficult to see the pupil, and to make observations on the reaction of the iris to light. Apart from this, the record of the development of vision will, it is hoped, be found pretty complete.

The "opening of the eyes" is really a separation of the lids, which are practically one at birth, by a process of growth and absorption along the line of their future edges. These processes take a few days for completion, even after there is an obvious opening between the lids; and it is very doubtful if the animal sees at all, in the proper sense of the word, until the lids are completely separated, if even then; so that the eyes being open is in itself no guarantee that the animal sees, or, at all events, more than light and shadows.

The slowness of reflex winking to appear in puppies is surprising, the more so as mature dogs wink very readily when any object is brought near, or moved before the eye.

Quite otherwise is it with mature birds, and it is almost impossible to get the young to wink, even on touching the lids in some cases I have found.

In this, and a former litter of St Bernards, the eyes began to open on the 11th day, and in a litter of Bedlington terriers on the same day, or perhaps a little earlier. One writer states that the eyes of dogs open on the 8th day. I have never seen this, and do not believe it holds for any pure-bred dogs, at all events.

But individual differences show to the extent of at least twelve hours.

HEARING.—It is very easy to be deceived in this, on account of motor effects resulting from concussion, or from contact of blasts of air with the skin. I think, however, my experiments will be deemed conclusive, and the record of the development of this sense very full.

There comes a time, as I have noted, when the young dog is more affected by sounds than an older one, owing to the less perfect development of his cerebral cortex, which part of the brain is associated with all higher psychic manifestations, with voluntary movements, inhibitions, etc. To this the lack of experience is to be added, for till the dog has learned better, noises of all kinds are excitements which may have unpleasant associations or the reverse. The mature dog has embedded in his nervous system and psychic nature a series of connections which, without any reasoning answer to warn him or the reverse, are perfectly indifferent.

However, new and mysterious sounds may alarm a mature dog more than a puppy.

The lower animals are more sensitive to concussions than man, as shown by their behaviour prior to earthquakes, when there are slight oscillations of the earth, wholly unperceived by man, yet causing alarm to the domestic animals.

I have noticed that puppies are very early stimulated by concussions, but regret that I have not exact observations with fixed dates to report.

One of the earliest indications of hearing is *reflex movement* of the ears. These are quite distinct, of course, from the voluntary movements often seen in dogs and other animals. But similar, though less marked, movements of the external ears may be observed in man also, as any one may prove by asking an individual to listen and determine the location of a tuning-fork sounded behind him. These I have for many years been accustomed to demonstrate to my classes in physiology, though I have not noticed that they are referred to in books. There seems to be no relation between the extent of the reflex and the voluntary movements of the ears, of which some people are capable. When at concerts I have sometimes observed them in great numbers and variety.

Another matter that seems to have received scant attention, if I may judge from the absence of printed references, is the condition of the ears in puppies up to a certain date. At birth the external ear is turned back, and its internal aspect strikes one by its relatively undifferentiated character, and the *auditory meatus* is scarcely to be recognised. The ear, in fact, grows and differentiates after birth in somewhat the same way as the eyelids, but the latter are invariably in advance, so that there are physical reasons for the deafness of puppies. Even after the ear seems to be opened up, the introduction of a fine probe is impossible, as I have shown.

Psychic manifestations may be looked at from so many different points of view, and the correct interpretation is so often doubtful, especially in the lower animals, one's explanations are apt to be so artificial, narrow, or otherwise imperfect, that I shall, under several headings, now refer to the early development of the puppy.

PLAY.—I have endeavoured to follow very closely the development of the play instinct, so important is it as a means of physical and psychic development, as well as an indication and an index of the latter—in fact of both. The reader is referred to records of the 13th, 15th, 21st, 27th, and 32nd days more especially. I have felt keenly my inability to record all that I have seen in this connection, not to mention the thoughts suggested, which lack of space prevents me making even an attempt to indicate.

What is play? One observes, first of all, that the puppy uses its mouth generally on a fellow, then, or simultaneously, its paws; but soon the movements are more complicated, prolonged, and accompanied by various vocal expressions, which are of a significance which varies with the age of the puppy.

There is not the slightest attempt at play during the period of eye-closure.

At first, playing seems to arise in part from an excess of motor energy which must be discharged, and as it is in the nature of the dog to use his jaws so much, the play takes the special form of biting; then the mouth is naturally assisted by the forelimbs. As locomotive power increases, the puppy takes to walking away and returning to the attack, then running, jumping, etc.

Soon he begins to shake objects, pull at them, tear them. My observations show conclusively that the movements in play appear in the order of the final perfection of the co-ordinated movements of the animal as represented, so far as the nervous system is concerned, in the cerebral cortex by well-defined centres. I am now, and for some time have been, engaged upon experiments which show that the cortical brain centres do not all develop at the same time, but in a certain order, a fact which throws a flood of light on the psychic, as well as the physical, development of animals.

The pleasure of play is that of movement at first. Later, there is no doubt a psychic complexity of feeling not known to the very young puppy.

Nevertheless, the observations reported on the 26th and 33rd days would seem to indicate that even at this early age the puppy has some sense of fun or humour.

SCRATCHING.—I have endeavoured to note the earliest attempts at this act, and give some details from time to time, as it illustrates several points.

I should be disposed to regard scratching as a hereditary reflex perhaps, as is illustrated by the experiment of the 17th day. In other cases, however, the element of will does enter more or less into this act. Even an adult dog will move his leg in the air in harmony with scratching irritation against his side—a pure reflex. When, as noted on the 40th day, the puppy turns his neck so as to adapt the movements of the leg, and the position of the parts to be scratched, it is plain that we have here the element of will as well as a fine example of neuro-muscular co-ordination.

The study of the development of such acts as scratching, and that next to be referred to, are very suggestive and instructive to the physiologist and psychologist. I call special attention to this reflex and its psychic effects referred to in the diary on the 17th day, and, in the case of the Bedlington terriers, on the 30th day.

WAGGING OF THE TAIL.—The tail movements of the dog are so expressive that the history of their development, and the analysis of their meaning at the various stages of the evolution of his life, are of more than ordinary interest. They are to him what words are to mankind.

It is notable that I have been unable to be positive as to the existence of any tail movements during the period when the eyes are unopened, and this alone is significant of the relatively low state of development at this period. The reader is referred to the records of the 13th, 17th, 21st, 23rd, 28th, 31st, 35th, and 42nd days especially for notes that bear on this subject.

These movements, positions, etc., of the tail have been to me signs of great significance, but I will leave the reader to draw his own conclusions. Certain it is, they are characteristic of certain stages of development, but if I were to go into full detail in reference to all they have suggested, this paper would become of inordinate length. It throws not a little light on this subject to remember that a centre for tail movements has been demonstrated in the cerebral cortex of the dog.

SOCIABILITY.—Of all animals known to us the dog is the most sociable. This he early indicates by his tail, the expression of his face, his attitudes, locomotive movements, voice, etc., and the reader is referred to the diary for evidences of a development of these characteristics of his nature, especially as regards man, a development which is so rapid, after the 30th to 40th day, that the puppy, in a few weeks, has become, in this respect, very like a mature dog.

FRIGHT.— The diary contains references to this subject on the 26th, 33rd, 37th, 39th, etc., days. After hearing is established, fright is easily caused through that sense, and apparently much more readily than through vision at a very early period. At this time also concussions, as such, are potent in producing fear. I regret that the influence of concussions was not more fully tested during the blind period. I find that the Bedlington terriers were thus alarmed on the 9th day.

Though the phenomena witnessed, when a puppy a day old is in danger of slipping off a surface of support, suggest alarm on its part, I question whether the puppy is possessed of enough consciousness, so to speak, to experience true fright.

VOICE.—Puppies may, and usually do, cry (in a manner scarcely to be distinguished from a kitten, so that mature dogs hearing it, bark, thinking cats are about) almost as soon as born. Gradually this voice is changed to that which is characteristic of the dog. Before barking in any form, growling in sleep, then in play, has been observed. They were heard to bark in sleep before doing so when awake. Such use of the voice is reflex or similar to reflex action.

The diary contains the earliest observed use of the voice in various ways with the circumstances stated, and, among others, I call attention to the records for the 23rd, 27th, 35th, 42nd, and 59th days.

It will be noticed again that there is no proper use of the voice beyond crying during the blind period, and that there is a development of growling in sleep, growling when awake, barking (incipient as in older dogs) during sleep, probably in dreams, barking simply as an expression of surplus of energy, barking in wantonness, etc., all of which is, like the wagging of the tail, highly characteristic of different psychic states.

All these modes of expressions are to be witnessed with precisely the same interpretation in older dogs at times, though, of course, generally the meaning of their barking and growling is more definite. But the puppy persists latent in the dog, just as does the boy in the man.

DREAMING.—Mature dogs do undoubtedly dream, and if one may judge by similar use of the voice and like general behaviour, puppies do also. Leaving out of the question the doubtful evidence of growling in sleep, the phenomena reported on the 35th day seem to point to dreaming, for the behaviour of the puppy is very similar to that of the mature dog.

ANGER.—Much of the play of dogs is mimic fighting, even from the first, and I have noted on the 27th day, during play, a very brief but decided exhibition of anger, such as may occasionally be seen among mature dogs, or boys even of eight or nine years of age during rough play. For the moment anger rules, and the extent to which this is the case, and especially the length of time over which it lasts, depends greatly on the breed of the dogs. With terriers very early play at times becomes serious, and later it may so often become fighting that these dogs cannot always with safety be left together. In few respects do the different breeds show their characteristics, or at so early an age, as in this. For a very early case of anger (or was it a mere reflex?) see the record of the 17th day, and for a clear case the record of the terriers on the 30th day.

MEMORY.—In a sense all impressions are remembered, *i.e.* the state of the nervous system, indeed the whole organism, somatic and psychic, is dependent on impressions, ancestral, pre-natal, and post-natal. It is simply impossible that it should be otherwise. However, in the more restricted sense of the word "memory," a good instance is to be noted in the behaviour of the puppy that was accidentally trodden upon by me. This occurred on the 47th day, and up to the date of the conclusion of the diary, on the 60th day, it was very clear that he remembered this unpleasant event.

Memory is very retentive in dogs, though there seem to be, in this respect, as much individual difference as in human beings. I had a greyhound that could not see a cat on the street without giving chase; and he would, after many months, remember the identical tree up which the cat climbed when he was in pursuit. This is, moreover, a case of visual memory in all probability, as it is not likely that the scent from the cat would remain for six months.

RECOGNITION.—From several experiments recorded, as the result of introducing other bitches into the same compartment with the puppies, the reader may be able to draw some conclusions. From the behaviour of the puppies I conclude that, at the time of the later experiments, the fact that they attempted to suck the strangers is not evidence that they were mistaken for the dam, but that they simply had such a desire to suck that they were willing to accept what they could get. They, in one instance, gave the clearest preference for the dam, and at once, guided probably by sight chiefly, for dogs' judgments are quickest by sight, though often corroborated by smell. Smell is their surest guide, and always called into use in doubtful cases. See especially the record for the 36th day. Of course, I witnessed evidence for my conclusion, which in this, and other cases, it is not possible for me to fully communicate by words

I have noticed in these and other puppies a quick recognition of human association through what I have termed the lip-call, not to be identified with any other sound. Is this the result of heredity to any extent, this sound having been used more than any other in attracting the attention of dogs? But so readily are psychic associations formed that one must not be sure of this. The dog, above all our domestic animals, is a plastic creature, and his life is made up largely of associative reflexes and kindred neuroses with corresponding psychoses. This principle I regard as a key that unlocks more of the secret places of canine nature than perhaps any other, unless it be heredity itself.

HUMOUR.—The records of the 26th and 33rd days seem to show that even such young puppies appreciate fun or humour, much as a child does, and this can be almost daily observed in mature dogs.

ATTENTION AND FATIGUE.—My observations on these subjects, some of which I have attempted to record, show, in the plainest way, how very readily a puppy is fatigued, but also indicate a gradual improvement in this respect. This readiness in experiencing fatigue explains why, moreover, one observer may be led to question the observations of another on very young animals. Again and again have I failed in my attempt to get the same result on repetition. In fact, the rule up to about the 20th day was, that success on repetition of certain stimuli was very doubtful owing to fatigue.

This is well illustrated in the case of the growling reflex, etc. of the 17th day, but it applies to all the senses and the whole life of the animal, somatic and psychic.

For this reason sleep follows at once on the exertion of play, with its physical movements and its sensory stimuli tending to exhaust.

L

Hence, too, the necessity of abundance of sleep in early life for all animals.

How important that this state of things should be recognised by all educators—in fact, all who have to do with young children, to whom it applies equally with dogs and other young animals!

CONSCIOUSNESS.—The dependence of consciousness on sensory impressions is readily shown. It was found that bandaging the eyes of the puppies sufficed, on the 18th day, to quiet them, and even put them asleep, when in their usual environment (pen).

This subject is evidently closely akin to the previous ones. While these relations exist all through life their clearest demonstration is in the young animal.

DREAMING.—If mature dogs dream—and of this there seems no reasonable doubt—the phenomena witnessed in the puppies on the 35th day is evidence of the same state. Growling in sleep was noted as early as the 17th day, but I would hesitate to refer this to dreaming—in fact, I do not think such an explanation applicable if the term "dreaming" be used in the same sense in which it would apply to a mature dog having a vision of imaginary events that arouse feelings.

WILL.—It may, perhaps, be doubted if there be any appreciable exercise of will proper during the period when the eyes are unopened; but on the 17th day, when on the puppy's ear being rubbed gently, he, in addition to scratching, puts up his foreleg occasionally, as if to remove the source of irritation, there is the appearance of volition. At first reflex and voluntary action are much mixed, of which there are many examples to be picked out from the diary, but in some instances cases of pure volition may be found, *e.g.* when on the 20th day the puppies go to the wall of their pen and attempt to get over it. But even this is to me by no means so clear a case as that of the 41st day, when a puppy watches a fly that has been tormenting him, and then, steadying his head deliberately, snaps at it like a mature dog.

SUGGESTIVE ACTIONS.—I prefer this term to "imitation," as the latter has become associated, in most minds, with the attempt to repeat what has been seen. In dogs the first imitative action, or rather suggestive action, is seen in play. One bites the other gently, and this rouses the tendency to reciprocate. It comes before all visual suggestive action. When several mature dogs are kept together, one may witness daily many interesting examples of imitative action. It has an educative effect of the widest influence either for good or evil on dogs. Much of sheep-worrying, etc., is the result of suggestive action, and is not spontaneous, except in so far as it is natural to all dogs to chase.

In the puppy, from the 40th day onward, suggestive action is very common, and this greatly increases the activity and hastens the psychic progress of the members of a litter of puppies, as compared with a single young dog kept apart.

It often, I have noticed, advances a puppy of a few months of age to place him among older dogs; and this is sometimes followed by the best physical, as well as psychic, results, especially if the young dog be allowed to go out for exercise with the older ones, under direction, of course, for dogs should not be allowed to roam as they will any more than children. They, too, soon learn the ways of the street. The manner in which this principle of suggestive action was illustrated on the 61st day, when in the yard among the older dogs, was very striking.

RESEMBLANCES TO THE MATURE DOG.—Every animal is what it is by reason of its inherent tendencies as reacted on by the environment, and at this stage it may be interesting and instructive to call attention to the first occasion on which actions suggestive of those of older dogs, if not practically identical, were manifested. The reader is especially referred to certain records on the 37th, 39th, 40th, 42nd, 43rd, 45th, 47th, 48th, 49th and 50th days.

Indeed, after the 50th day, these resemblances in behaviour are so numerous, or, in other words, the puppy is so matured, so fully equipped psychically, that much less interest, or at all events importance, attaches to the study of his psychic life.

INFLUENCES OF ENVIRONMENT.—As has been explained, when in the young puppy the eyes are closed, he is very apt to fall asleep, and if all the stimuli through the sensory organs were cut off, consciousness would be reduced to a minimum, if it existed at all. On the other hand, as illustrating the influence of the environment, in special ways, on the early psychic life of the puppy, the reader is referred to records in the diary on the 23rd, 26th, 33rd, 45th, 46th, 47th, and 49th days among others. There is not space for comment.

REASONING.—I do not propose to enter into the controversy as to whether animals not possessed of articulate language can reason, or whether we should name the process corresponding to that in man, "inference."

That man can reason in a way that animals lower in the scale cannot, is certain, but that much that we assume to be of a higher order in the mind of man, and perhaps consider reasoning of this higher order, differs in no essential point from psychic processes in animals, I am convinced, after many years' close observation alike of animals and man, including the working of my own mind, which, after all, is the final court of appeal for oneself. When, on the 41st day, the puppy scrapes away the sawdust, and then some days later, repeating the act, tries one spot with the head, not being satisfied paws again just where there is a slight elevation in the floor, is there reasoning?

When on nearly every occasion on seeing me, the puppy that had been trodden on retired with his tail down, and an appearance of dejection, did he reason that I might be again the cause of some unpleasant feelings to him?

Two evenings since, the weather being intensely hot, the dam of these puppies was allowed to sleep on a veranda (more airy) of the house instead of in the kennel. She had not been on this veranda since last summer. At a late hour I opened the door leading from the veranda into the yard, and invited her to come out. She declined to do so, which at first surprised me. The dog did not wish to be removed to the kennel, and this was borne out by the fact that on the following evening, as she lay on the same veranda, opening the door leading to the yard, and at the same time that of the kitchen, she immediately got up and walked into the kitchen. In the latter she had received many a tit-bit. Wherein does the behaviour of this St Bernard bitch differ from that of a child of, say, five years of age who, when amid his play, is called by his mother, but silently protesting turns quickly away? Does he, before turning, formulate any sentences? He can do so, to be sure, but does he—must he? Is not the process, or series of processes, in his mind closely akin to those in the mind of my St Bernard?

Is the behaviour of the puppy that turns away when he sees me different from, or akin to, that of its dam, in the circumstances already detailed?

In the case of pawing away the sawdust there seems to be the recognition of a cause, yet it is possible to separate this mental process wholly from the restless moving about of an animal that does not find its bed quite comfortable, and which certainly requires no "reasoning" to explain?

ASSOCIATED REFLEXES.—When referring to reflexes in general, I omitted to call attention to certain phenomena which seem to me unquestionably of this character, *e.g.* on the 23rd day, when one puppy licks the other after feeding, as is always the case, it is observed to place its paws on the head of the other, and spread the toes exactly as in sucking the mother, when it places its paws against the mammary glands, and so in other instances. The association in one kind of use of the mouth (sucking) is made with another kind as licking, etc.

EXPERIENCE.—Any one who without prejudice watches any young animal, cannot fail to be impressed with the readiness with which, within certain natural limits, it profits by its experience; and this is one of the lessons of the diary of these puppies, evident in all directions, instinct included. As one instance, among many, I refer the reader to the advance noted in regard to the bones on the 59th and 61st days, and the entire behaviour of the puppies in the yard, on that day. The manner in which they acted, as if they were well acquainted with the yard, the various ways in which their movements and actions suggested the old dog, illustrated to me in a way, that was somewhat of a surprise, the readiness with which they availed themselves of every experience, and quickly worked it into their nature.

THE MYSTERIOUS.—That dogs do, in some fashion, recognise causation, and are puzzled by its apparent absence, seems to be beyond doubt.

The earliest manifestation of this I have noted on the 38th day, in connection with dangling a bright chain; nevertheless, this is not to my mind a clear case.

INDIVIDUALITY.—From time to time reference has been made to individual differences, both psychic and physical. It is not easy to make perfectly evident in a diary the extent to which individuality is shown, but even in the blind period it exists, and to a close observer, familiar with dogs, and the particular breed being studied, it shows itself in a variety of ways, often it may be difficult to describe in words. Sometimes, when but a few weeks old, a puppy foreshadows his future in an unmistakable way.

PERIODS OF DEVELOPMENT.—A study of the diary will show that the two great periods are: that before the eyes are open, and that succeeding this one. The time between the opening of the eyes, and the establishment of real vision and hearing, constitutes a transition or intermediate period.

Development is very slow in the first period, and existence almost a vegetative one, yet not wholly so, for by the skin, the muscular sense, to some extent, by taste and smell, by visceral sensations, etc., the animal's nervous centres are being modified.

The intermediate period is marked by a considerable advance, though slow, as compared with the progress made within the next few days.

The period between about the 17th and the 45th day is that of greatest importance in the life of the dog.

After that there is constant improvement, from experience, up to the 60th day, and this is well marked—more so than at any later time, but it is not of equal importance with that preceding.

These periods glide into one another, and many others might be interpolated, but I desire to avoid artificiality, which is sure to result from the attempt at numerous divisions of any kind.

There is not the sharp line of difference between the dog and other animals, before the eyes are opened and afterwards, which some writers would have us believe, though between the animal, when it can neither see nor hear, and the same animal ten days afterwards, there is indeed a vast difference. But as to the rate and nature of development the reader may draw his own conclusions, and to enable him to do so has been my chief object in giving a record of facts so detailed and as free from gaps and omissions as possible. I am convinced, moreover, that the whole difference in the periods referred to is not to be referred merely to the presence or absence of vision and hearing.

About this time the whole nature of the animal seems to undergo a comparatively sudden leap forward in advancement, possibly as the result of the accumulated experiences of ages acting through heredity—I mean that the advances directly referable to the advent of seeing and hearing would tend to accumulate by heredity, and to be expressed in the organism in time in a more decided manner.

GENERAL.—The preceding are a few of the many aspects of the psychic (and physical) development presented within the first sixty days of existence of puppies. I deprecate hard and fast lines of demarcation in biology and psychology, believing that in nature one thing, as a rule, glides into another at some stage of development, at all events. My commentary on the diary is, therefore, not claimed to be complete, if indeed it is possible to recognise all that there is in psychic development, however closely one may observe, however perfectly analyse.

PHYSICAL CORRELATION.—Already, for some years, the relations of mind and body have been recognised in a general way, and studied with results of definite value; but while there have been isolated experiments and observations made on young animals bearing on the relation between physical development and the psychic status, I am not aware that any complete and systematic study of the subject has been attempted. That the mind and the body must be studied together will, I am satisfied, become more and more evident as investigations on the one, independently of the other, prove disappointing. This applies more particularly, no doubt, to the mind, but not wholly. While to a practised observer very many shades of change in physical developments may be observed, there is no good method of measuring most of them, and it is more than difficult to express much of what is observed in a way to make it appreciable by the mind of the reader.

Until our knowledge of the relations between the mind and the body—between the history of the body and that of the mind—between ontogeny and psychogeny (psychogenesis) is made very much more complete, it would appear that it is desirable that a contemporaneous account be kept of every change of whatever kind observed, both physical and psychic.

We dare scarcely say that matters so apparently trivial as the change in colour of the iris, or as the pigmentation of the nose, for example, are in no relation whatever with psychic development.

Has the eruption of the teeth in the puppy no relation to psychic growth and development? In itself the direct causal relation from increasing experience thus afforded by their use is not all, and there is doubtless in this more than we are in a position to define as yet. As soon as the teeth appear, and the jaws are more used, as is now the tendency, the puppy advances in consequence of this very use of teeth and jaws, but this is probably not the whole story.

From the chief diary, and the comments on it, the reader will be able to cull many instances of psychic and physical correlations. Between the physical changes in the eye and ear especially, and the psychic results, the closest relation is evident, and this should suggest that similar close connection exists elsewhere,

While the puppy sprawls in the blind period, he cannot investigate objects, and we find, as the sensory organs advance in development, the animal's locomotor power increases, so that he can the better use all his senses, hence the great strides he makes in development from one part undergoing a change which adapts it to the well-being of other parts and the entire organism.

As a matter of fact, motor power is, in the young animal, a very fair guide to its general advancement, and in tracing the development of the puppy one notices this daily.

There is a certain order of progress: first the tongue laps, etc., as in sucking; then after the eruption of teeth, use of the jaws at the same time; and more so later the movements of the fore-limb—long, in fact always, in advance of the hind-limb—the tail soon taking a share in the movements.

These movements not only increase in power, but in precision, *i.e.* they are co-ordinated, and this is well illustrated by many facts stated in the diary.

These movements, the development of the senses, etc., etc., are of course impossible without the nervous system, and they gain in precision and variety, according to the rate and extent to which the cortex of the cerebrum is developed into functional activity. My own experiments on the brains of young animals are not yet complete, so that I shall not here refer to them further than to state that they bear out the view just stated. During the blind period the cerebral cortex is found to be unexcitable, while in the mature dog movements of definite groups of muscles may be readily obtained by stimulation of the cortex.

DIFFERENCES IN BREEDS. — Both physically and psychically there are differences in development in the various breeds of dogs.

I found that the litter of Bedlington terriers

developed much faster psychically than St Bernards, and they also mature earlier, physically and otherwise, a remark that applies to the smaller breeds of dogs generally.

They sooner show, especially in movements, a great superiority, which strengthens the opinion I have expressed, that, among animals, the degree of advancement in co-ordinated movements is a fairly good guide to psychic progress at early periods.

WHAT REMAINS TO BE DONE.—I am now anxious, as all my work has been done on pure-bred dogs, to study a litter of mongrels.

It has been thought well to confine this paper to the study of the early development of dogs.

I can see the desirability of supplementing this paper by the account of some one dog from birth to maturity, and possibly I may be able to do this.

I purpose following this paper by another similar one on the development of other animals in the earlier periods of existence, considerable material for which has already been accumulated, so that I hope in time to get the facts in such form, that broad and sound conclusions as to development of young animals may be drawn.

As the dog, after the monkeys and apes, more closely resembles man psychically than any other animal, it seems to me that it would be very profitable to attempt a comparison of the development of the young dog and the infant. But this task must also be deferred.

For various reasons I have not referred in detail to the fragmentary work of others, chiefly because the original papers are not, in most cases, accessible to me now, and because prolonged discussions and comparisons with their results would add to the length of an already long paper. I present my observations with such conclusions as I have tried to draw cautiously, and without prejudice, believing that, whatever their defects, they constitute the most complete account of the subject, published to date.

SOME CONCLUSIONS.—The dog is born blind and deaf. He possibly smells and tastes feebly, but this is difficult of demonstration; but in any case he smells, tastes, has tactile and muscular sensations, the temperature sense, and can experience pain before he can either see or hear.

The eyes are open before the ears, but seeing objects does not correspond in time with the opening of the eyelids, which is gradual, the result of processes of growth and absorption. Hearing follows sooner on complete opening of the ears than seeing on opening of the eyes.

There is progressive improvement in both seeing and hearing.

Both begin about the 17th day, and are in a high state of perfection about the 30th day, hearing being, upon the whole, rather more rapid in development.

Smell and taste are demonstrable on the 13th day, and are well developed about the 30th day.

Newly born dogs are very much affected unfavourably by a temperature below a certain moderate point (50° or 60° F.), and are capable from the first of such movements as enable them to avail themselves of the heat from the mother's body.

They give evidence of feeling hunger, and are capable of making certain slow movements at birth.

They find the teats chiefly, if not wholly, by touch, and continue sucking in consequence of the satisfaction of the appetite for food.

Up to about the 20th day puppies are very readily fatigued, and incapable of attention to anything for more than a very few seconds at one time.

They early show an appreciation of any decided

change in the environment, indicating that experience, even in the earliest days, is not lost on them. In other words, the environment does and must act on the nervous system, with results that manifest themselves if in no more definite way, at least in this: that new experiences (stimuli) cause comfort or discomfort, as evidenced by quiescence or wriggling, cries, etc.

Co-ordinated muscular movements appear in greatest perfection in a certain order, viz. mouth and head parts, fore-limbs, hind-limbs, tail, etc.

These seem to be related to the order of development of the centres of the cerebral cortex.

The epochs most differentiated from each other in the psychic and somatic life of the dog are (1) that prior to the opening of the eyes, and (2) that subsequent to this event.

The former suggests intra-uterine life by its negative character, and is well marked off from the period that follows, the more numerous avenues of knowledge existing, and their utilisation, and in other respects not well understood, of the latter period. In other words, the animal, after this period, can come more fully in contact with environment, with corresponding results in its development. It seems, besides, more impelled to do so; there is more vim in its whole nature. A transition period between the time when the eyes and ears begin to open, and when the animal actually sees objects and hear sounds, may also be recognised.

The era of most rapid and most important development is subsequent to the period when seeing and hearing are established—when the animal is in possession of all its senses, etc. This extends between the 20th and the 45th day approximately.

Suggestive action, beginning perhaps with the first manifestations of the play instinct, has, especially as time passes, a very important share in determining the direction of development, and what manner of dog the individual becomes. It is education in the more limited sense.

The order of development of the senses and co-ordinated movements as well as reflexes, and the manifestation and perfecting of instincts, have a distinct relation to the needs, as well as the general development of the animal, *e.g.* smell is always more important to the dog than any of his other senses, and it is early developed. The same remark applies to the movements of the jaws and the limbs over those of other parts.

The detailed study of the development of the dog, as recorded in the foregoing pages, illustrates how dependent all subsequent advancement is on the early and full development of the senses and co-ordinated movements. They bring the nervous centres into contact, so to speak, with the environment.

The same is illustrated in the study of the human infant; but in the case of the dog the investigation is not surrounded by the same complications or, at all events, prejudices.

Although it is not possible as yet to determine the physical and psychic correlations down to the minutest details, from what has been accomplished, it seems reasonable to hope that a complete correlation may be ultimately established.

The first sixty days of a dog's existence are of so much more consequence than any later period, that the writer has decided to limit this paper to this period, within which almost all important features in development appear.

II.—The Cat.

THE present paper is a continuation of that series on the psychic development of animals, or psychogenesis, the first part of which appeared in the *Transactions of the Royal Society of Canada* for 1894.

As the desirability, purpose, and scope of such investigations have been set forth in the paper on the dog, just referred to, no lengthened introduction will be necessary in the present instance.

The records were made under more favourable circumstances, and are more extensive and complete than those concerning any other animal that I have, up to the present, been able to study.

The kitten seems to me to have been one of more than ordinary interest, and though the observations extend over 135 days, had the animal not disappeared, I should have continued my records.

The diary will tell to each reader its own story, and I shall therefore make the observations upon it somewhat brief and suggestive, rather than attempt to exhaust the lessons it teaches, and as this paper will be followed by one in which the dog and the cat will be compared, there is additional reason for making the notes upon the records, and the part of the paper devoted to conclusions from the observations, briefer than they might otherwise be.

The readers who peruse the first of the series, and the subsequent papers, will naturally derive more profit—whatever that may be—from the present paper.

Diary.

The following notes were made on a litter of kittens, the parents of which were ordinary domestic cats.

They were born 29th July 1894, and came first under my personal observation two days later.

2nd day.—Eyes and ears closed. No evidence of hearing on sounding a shrill dog-whistle, etc.

Rubbed hands well on a St Bernard and a beagle dog and placed them near the nose of one of the kittens. It *sniffed* and became *uneasy*, but did not hiss.

The beagle was induced to lie down and the kitten placed against its belly (male dog). The kitten *turned* away.

With a view to testing *taste*, salt, sugar, and aloes in solution, as well as cow's milk, are used. Kitten sucks a feather dipped in solution of sugar, but manifests no sign of disgust when aloes is substituted. There is uncertainty as to salt and milk.

As to *smell*, aloes and iodoform brought near the nose cause the kitten to open the mouth and show signs of *disgust*. Blistering fluid (liquor epispasticus) and strong carbolic acid produce the same effects in a minor degree.

When milk is presented no attempt is made to lap it.

When I try to introduce milk into the mouth with my finger, the kitten uses its fore-paws to remove finger.

At present, the animal may be said to *crawl* rather than walk. It shows great *uneasiness* when it gets near the edge of a table, and *holds* on vigorously with its claws, manifesting uneasiness by its cries. Removal from its ordinary environment of comfort also results in crying.

When the end of a glass pestle, heated to a degree uncomfortable to the human skin, is placed against the sole of the kitten's foot, *withdrawal* follows.

Gentle *touching* of the mouth or nose, especially the inner surface of the nostrils, leads to a decided retraction of the head.

It seems to require a good sharp pinch to cause withdrawal of a part in a way to indicate pain, and the latent period is notably long.

6th day.—Evidence that iodoform is perceived at a ½ inch, blistering fluid at a ½ inch, carbolic acid at 1 inch, and aloes at 2 inches.

The kitten manifests a desire to escape from salt and aloes but not from sugar.

Moves better; makes an attempt to crawl out of a small basket which is lying half tilted over.

The same sort of a pinch as before causes more decided movement with a shorter *latent* period.

The kitten shows no less uneasiness, but rather more, when brought near the edge of the table.

At a distance of 4 inches from the beagle dog the kitten *opens* its *mouth* and spits. When the *hands* are rubbed on the beagle, and brought near the nose of the kitten, the same thing happens, but the hands must be brought within 1 inch of the nose; and the animal seems to mind it less and less, though this is not so manifest when brought near the dog. It still turns away from the belly of the dog when laid against it.

8th day.—Eyes begin *to open* in two of the three kittens.

When I make a loud hissing noise the kitten with the open eyes opens its mouth and hisses, and, when the loudness of the noise is suddenly increased, its jaws snap together audibly.

The use of the dog-whistle is followed by uneasy but not sudden movements of all three kittens. No sudden movements of the ears in reaction to sound at present, though they evidently *hear*.

9th day.—The eyes of one of the kittens still not open.

Two of the kittens tested seem to smell equally well iodoform at ¾ inch, carbolic acid at 1 inch, blistering fluid at 1½ inch. *Dislike* manifested by turning the head away and putting up the paw to nose, as if to push away something. The mouth is also opened.

Kitten *licks* with evident pleasure at some candy. It licks at salt also, but soon shows disgust. Aloes give a doubtful reaction.

When the foot is pinched the kitten now *cries* out sharply.

There is a noticeable improvement in locomotor action.

It is impossible for me to convince myself that any of the kittens see. There is a strong tendency to keep the *eyes closed* a good part of the time. The *winking reflex* is not produced by moving objects before the eyes, but follows only when the hairs of the lids are reached, or the lids themselves, or some other part of the face or the head is touched.

On the other hand, both a sharp sound from the whistle and a loud sound, imitative of a bark, causes the kittens to start, but a *hissing* sound affects them most, and at a distance of 3 to 4 feet will cause them to open the mouth and hiss.

The *ears* now move reflexly to sound, but turn towards the point from which sound comes, or, at all events, towards the side rather than straight back, as in dogs, or as in rabbits, etc.

Quivering of the ears is noticed, but the cause is not obvious. They turn to the quarter from which sound comes.

10*th day.*—No clear evidence of power to distinguish objects by the eyes as yet. The ears are noticed to move without any apparent stimulus. The quivering still continues. With each mew the kitten shuts its eyes.

A sharp *whistle* causes the animal to *start*, but the ears move but little. The kittens crawl rather than walk yet, though the pace is increased. There is some amount of *tail* movement.

11*th day.*—Two of the kittens give evidence of seeing, though one rather better than the other. The eyes are now well open. Seem to *see* at 10 to 12 inches' distance, though the evidence of this is not conclusive. *Winking reflex* on movement of the fingers before the eye at a ½ inch or a little more.

They can now cry with a relatively loud *voice*.

On testing taste with aloes, *disgust* is very plainly shown.

Ears seem to be moved *voluntarily*. The quivering continues.

12*th day.*—When cooked beefsteak is held within 1 inch of the nose it is *smelt*, as evidenced by sniffing, etc.

When held near the nose of another it hisses. Neither shows any desire to secure the morsel. When placed in the mouth of one of them it is allowed to drop out.

The *pupil* is now observed to act, and there is intolerance of strong light. The eyes, however, are not quite free from irritation.

Very sensitive to *touching* by hand anywhere about the head. This was noticed in attempting to cut, ever so gently, hairs on the head with scissors.

Kitten *turns* to right or left as sound comes from one side or the other. Sometimes *hisses*.

Still walks badly with *hind* legs.

13*th day.*—From this date only *one* kitten observed.

Is disturbed by iodoform at 2 inches from nose. The impression seems to be somewhat lasting. When aloes is within ¾ inch of the nose the kitten *hisses*.

Pupils vary readily with the amount of light.

Winking reflex as before.

It appears as if it moves its *own head* in response to

movements of the head or body of an observer at 2 to 3 feet, but there is some doubt about this.

When lifted from its basket and placed on a table, the creature manifests uneasiness and cries, as has been the case from the first.

Now walks a little better.

14th day.—When a small piece of cooked liver is held near the nose it *licks* its lips. When the liver or milk is put against the tongue there is no evidence of taste.

Now follows the *finger* at 10 to 12 inches.

15th day.—A small dog goes up to the basket in which a kitten is lying. It *hisses* owing to smelling the dog (not seeing). When out of the basket does the same at 4 to 6 inches.

Walks in about a three-quarters erect position.

16th day.—Is seen for the first time to *lick* its paw.

The kitten and its mother brought into the house, and placed in a box about 16 inches long, 12 broad, and 9 high, made comfortable by a flannel covering on the bottom. This box was placed within a few feet of where I usually sit in my study, so that observations were easily and frequently made.

Kitten uses fore-paws when sucking to press on mammary glands of mother. It often sleeps apart from the old cat now. Its growth begins to be rapid.

Is seen to use right hind-leg to *scratch* ear and head on several occasions, and before stopping makes similar movements without touching body.

On holding a small object within 9 inches or less in the box it *hisses* feebly. Is seen to be biting gently at the lower woollen covering in the bottom of the box. This is the first indication of *play*, or something closely akin to it, that has been noticed.

17th day.—Kitten follows with its eyes a small object at 2 feet distant, and later, my copy of a journal

(11 by 8½ inches) at 4 feet. Is able to *locate* whether a light "Hist!" is uttered on the right or left.

Licks its paws occasionally.

Sleeps a great deal.

18*th day.*—Seems to be *following* with its eyes the movements of a fly at 4 to 6 inches.

Turns to bite at objects and puts paws together—a sort of feeble attempt at play.

Climbs up side of box to the top and then cries, as though desiring to get out.

On calling "Puss! puss!" softly, at about 3 feet, the kitten turns toward the sound and moves the ear nearest to it.

A piece of cooked meat is held near, but no special effect is noticed.

At 5.15 P.M. the kitten is seen to be plainly following *flies* with the eyes at 18 inches, and slightly raises *paw* in an attempt to catch them.

Licks its own paw, and then licks mother's leg which is close to it.

In the *evening* it gets lively and makes attempts at play, and crawls over the mother, biting at her and itself (play).

The nose is now getting pigmented.

19*th day.*—At 11 P.M. to-day follows finger by lamplight at 4 to 5 feet, climbs upon mother, and sleeps there.

20*th day.*—When "Puss! puss!" is called to-day, the kitten *hisses* softly (surprise).

There is now nearly always a distinct movement of the ear towards the direction of sounds.

Is taken from the box to-day, and, when free, *walks towards* it.

21*st day.*—Grows well, still *sleeps* much, and is very quiet.

After being out of box a short time it *returns* to it

and tries to *climb* in. Makes two such attempts and almost succeeded.

Can now follow a finger at 6 feet.

Licks its right hind-leg, and holds it in a position convenient for the purpose.

Some *incisor* teeth have appeared both above and below.

22*nd day.*—Distinguishes a finger at 6 feet.

When I scraped the fingers against the box in which it lies, *hisses* (surprise).

Licks herself on neck and chest, a difficult muscular feat.

Biting skin like mother.

Now walks relatively well.

On two occasions to-day climbs to top of box from the outside, does not go in, but puts paws on inside.

Does not now return to box so soon when allowed out. It walks around, *smells* at spots on the carpet, but wishes to get into the box in about two minutes, and manages to *scramble* in with difficulty.

It is noticed that the paws spread greatly in walking, so that it may be said to be plantigrade at this period.

Between 6 and 7 P.M., for the last three days, the kitten plays a *little*, lying on its back and using mouth and paws.

A small dog is brought near where it stands on the floor. It puts up its *back* and *tail, hisses*—even *spits*—with the fur erected.

Can now follow with the eyes the journal before mentioned at a distance of 8 feet.

Its hearing is also plainly more acute.

Several times to-day it has stretched up in box, and looked around as far as possible. When allowed out it goes around investigating with nose and eyes, but still soon wants to get back. After failing once or twice, manages to scramble in, but without hurting itself.

Later the same day succeeds on the first attempt.

The method of descent involves not only considerable control of muscular movements, but some knowledge acquired by experience.

24th day.—Mother away. Kitten cries, as if wanting to get out of box. Is lifted out, and after walking around a good deal, climbs back in the same manner as before, but with much less trouble. Finding the mother there it begins to play with her. Its attitudes, etc., much more distinctly cat-like than before.

Takes notice of *shadows* in box and elsewhere at night.

25th day.—Takes no notice of milk placed in saucer before it, though held close to the nose, but when put on its lips is removed with the tongue.

A string dangled before it produces no effect.

Is noticed to *watch* its mother's actions more intently, as well as what in general may be going on around it.

When asleep I call fairly loudly, at a distance of 3 feet, "Puss! puss!" but with negative result.

At 11 P.M. desires to get out of box. *Hesitates* when on the edge, and finally *glides* down near the leg of a table, within a few inches of the box.

On stroking its head, it *presses up* against the hand like an old cat.

26th day.—Is playing in box early this forenoon. Now and then walks with *tail up*. Cannot always walk without some unsteadiness.

Notices spots on carpet; sinks *claws* into it.

Seeks a *corner*, cries, evacuates fæces, which the mother at once removed, as when her progeny kept closely to the box.

Later it leaves the box again, and is taken back by the mother, which carried the kitten by the loose skin of the neck.

Soon it leaves the box again and goes to a part of the

room where there are some book-shelves, the lower ones of which are not completely filled with books, but hold other things.

The mother follows it. The kitten is put back into its box.

First noticed to-day what seemed to be purely *voluntary* movements of the ears.

Continues to notice *shadows*, and to-night tries to put its paw on them.

Still takes no notice of meat.

Licks hind foot, and, while doing so, spreads its *toes* apart, as an old cat may be seen to do under similar circumstances.

27th day.—On getting out of box starts on a little *run* for the book-shelves.

It was taken from among the objects on the shelf, turned towards the box, and given a few taps. It ran on to the box and got into it.

The mother is in the box moving her tail to and fro. The kitten plays with it.

I suddenly appear near the box with a coat on—light in colour, with pronounced vertical stripes—when the kitten at once *opens* its mouth, and, on my going nearer, *hisses*.

Later is seen watching flies, at a distance of 6 to 7 feet, flying.

Now gives evidence of hearing slight sounds at some distance in the room, and apparently notices the notes of a piano downstairs.

The movements of the ears are more pronounced.

The kitten continues to show a strong desire to get to the book-shelves, about which the mother manifests some anxiety, which she evinces by staying close to her.

Mother is making her toilet—the kitten bites at her in *opposition*. The mother then seized her by the neck, and after holding her quiet for a few seconds, goes on with her operations, but puts one leg over her as if to be on guard against any further interference.

28th day. — *Feeding* again attempted. A small quantity of milk is given in a little saucer, which the kitten licks, but seems to get some of the milk into the nose, which causes *sneezing*.

Plays with a small piece of coal found in her box.

On suddenly throwing aside a curtain that covers the book-shelves, the kitten is found there, and *hisses*.

I place her in a rocking-chair, over 18 inches from the floor, but she manages to *scramble* down without injury to herself.

About 5 P.M. the entrance to the book-shelf was barred up. The kitten first tries every part of the barricade, then pushes in the curtain, cries with vexation, climbs upon a box near, leaps from this on the curtain, holding on with the claws. After trying again and again desists, and after a few moments returns to the attack. At last she gives up, returns to her box, settles down and sucks her mother, and then soon after falls asleep. Her attempts to get into this shelf and accomplish her object were to me a study of unusual interest, especially as the animal was only 28 days old.

The eyes are now capable of much quicker movements than before.

Now hears "Puss!" however lightly uttered, also hears dogs barking in the yard.

This has been a day of activity and notably great advancement.

29th day.—Quieter to-day (reaction ?).

Without its mother last night for the first time.

Takes a little milk twice to-day.

There is some difficulty in keeping the milk out of the nose, which is owing to inability to *hold* the head just right. When solid food, as *meat*, is put in the mouth, it is rejected.

Shows a desire to use *claws;* sticks them into objects.

In play have seen an incipient run on two or three occasions.

When the mother has been cleaning the kitten, after answering nature's calls, there have for some days been cries, but to-day there is active protest with teeth and claws, which, however, the mother heeds but little.

When placed on a window ledge, 2 feet from the floor (and seeming more), the kitten looks down and plainly would like to get down, but does not venture to try the descent.

After lapping milk to-day it cleans itself (toilet).

Is seen for the first time to *wash* its own face.

30th day.—To-day, a *pan*, containing some fine sand in the bottom, is set near the kitten's box which is to be used as a retiring place to encourage cleanly habits. The sand will be frequently renewed.

The studies in connection with this have proved very interesting and instructive to the writer.

The kitten climbs out of box, and goes to the *corner* of the room near the box, and cries. It is then lifted into the pan and soon passes urine.

It then visits the book-shelf and tries to get in; *cries*. Not succeeding, it returns towards its box, and having to pass the pan on the way, it puts one paw into the sand, but withdraws it and goes on.

It soon enters its box and sleeps.

Later, is out of its box.

The tearing of the paper wrappers from a journal startles it considerably.

In the *evening* leaves its box and plays a good deal. It can now walk well, and run in some fashion.

It carries *tail* in all ways now, and moves it more or less.

It makes many attempts to get into the book-shelf and at last succeeds.

For its play, even the leg of the table suffices. It *darts back* as if it was afraid of it, etc.

31*st day.*—Is out of its box; *cries;* is put into pan; after some delay urinates.

Plays with carpet, a piece of coal, the edge of box, etc., its *back* being arched, *tail* up, etc.

Notice movements of tongue, jaws, etc., when asleep, as if sucking or lapping milk.

This morning the kitten *stretches*, like an old cat, four or five times when it gets out of its box (first time seen).

A string, curled up at end, is dangled before it in the box. The kitten *starts* back and *hisses.*

In the evening it is found behind the barricade of the book-shelf sleeping on some books. It is taken out, but works its way back again. It finds getting out difficult, but perseveres.

To-day *plays* with a ball and a spool.

This evening laps milk *without* any trouble for the first time.

32*nd day.*—Kitten seems uneasy; is placed in the sand pan; after *crying* and *pawing* it passes urine. Then slips out and stretches itself.

Still some trouble with nose when lapping milk.

Enters the pan of its own accord. No results.

It tries the book-shelf barricade, but not succeeding gives up, and sits in its box near by, and grooms itself well.

Later, it makes a more *determined attempt* on the barricade, and with success. It has difficulty in getting out, but soon goes in again and remains from half to three-quarters of an hour.

Enters pan; cries; urinates; much licking of the paws after it leaves the pan.

On moving one of my fingers before it, the kitten boxes with it.

Physical.—Its *eyes* are changing colour, being a sort of blue-grey. *Paws* no longer splayed. Its *fur* is darker in colour. It is rather a light tabby. *Fore-legs* not quite straight. *Tail* carried vertically a great deal. To-day spends much time sleeping behind the barricade. When taken out, it goes back persistently.

Still sucks the mother, using the fore-paws vigorously in *pressing* on the mammary glands.

It may be said to *jump* out of the box now rather than scramble out, *i.e.* lands on floor with a sort of leap, though it still scrambles in much as before.

33rd day.—Found behind the barricade this morning before daylight. When trying to prevent its advance in a certain direction, the kitten evades me by running under a rocking-chair, where it is partly hidden.

See it trying to *catch* a fly in the pan to-day.

When put in the pan now it nearly always *paws* the sand whether there is anything further or not.

Mother and kitten play together to-day for the first time.

Though the book-shelves were closed by a curtain tacked on them, the kitten managed to get in, though I do not know how.

Now *goes towards* persons sitting or walking in the room, and wanders about more or less.

Is startled by water running from a sink with down suction to-day, though about 30 feet distant.

34th day.—In a sort of *perverse play* bothers its mother so much that she puts her forelegs about the kitten to hold it quiet. Then the latter scratches so vigorously with hind-legs that the mother cries out.

To-day the kitten is kept in the pan, notwithstanding restlessness and crying, till it *passes fæces* for the first time here. This followed by much grooming of itself, as in an old cat.

The kitten has now a look of much greater *intelligence*.

35th day.—Growing well. Weighs 1¼ lbs. Is given milk, then placed in pan and uses it.

To-day when called "Puss! puss!" as it lies behind the barricade, it comes out.

Now manifests pleasure in being stroked.

Its vision is now so keen it can rapidly follow movements of the fingers, etc.

It scrambles into the book-shelf by a new way and at a much greater height.

Is seen to *catch* at its own tail for the first time.

Two Skye terrier puppies are brought near the kitten. It makes a great fuss, showing all the signs of *anger,* etc.

Some cooked *fish* (of which cats are notoriously fond) was placed before the kitten. It licked this but did not eat any, though there seemed to be a certain amount of surprise and excitement.

36th day.—The day is dull, and the kitten, lies in its box a great deal in the forenoon, though later it rouses itself and plays with its mother.

I try for some reason to intercept the kitten when it *makes a long run* to escape.

To-day, for the first time, it *crosses the threshold* of the room door, but was scared back.

For the first time, too, it sinks its claws into *an upright object,* as an old cat often does.

Some raw meat is placed before its nose. It *sniffs* at this, but makes no attempt to eat it.

It begins to understand better the meaning of the call "Puss! puss!"

Its manner of play now much more active and complicated.

Is noticed *stalking* on two occasions to-day—once towards the mother's moving tail.

37th day.—This morning early the kitten enters the pan, of its *own accord,* and functions.

When I cough at 6 feet the kitten puts up its back (surprise).

It is given milk on the window ledge referred to before, and, when it has finished lapping, *scrambles down* between the wall and a rocking-chair.

Later it comes over to me and *climbs up my clothing* into my lap.

Visits the pan again of its own accord, and functions, after which it retires to its box and sleeps.

Climbs into a waste-paper *basket*, and, after moving about among the papers a little while, climbs out without upsetting the basket.

The kitten walks out of the room, but on being turned back towards the door, and the hands clapped, it runs in again.

Again uses pan of its own accord.

Walks out of the room and begins to *climb the stairs* leading to a higher flat.

Climbs into my wife's lap in the evening, and from thence *gets upon a table* beside which she is sitting, and plays with objects on the table.

At the sound of footsteps in the room inclined to retreat from the table.

38th day.—Slept long behind the barricade this morning. On awakening goes spontaneously to the pan.

To-day goes behind books on another shelf near my table, and, when removed, returns.

Cooked meat is *licked*, but not chewed or swallowed.

Goes from my daughter's lap to the table top and sleeps there. On awakening, it looks down after moving about some, crouches, but hesitates to go further.

About 9 P.M., on this day, when sitting on one of my legs, the other being separated from this only about the length of the kitten, wishing to get across to the other leg, it fears to leap, but walks up higher, where there it finds a natural bridge.

The kitten is put on the window ledge where there is some milk. It soon wishes to get down, hesitates, cries, looks to mother apparently for aid, walks along the whole ledge, in doing which, as the window is open, it must pass over a surface only 1½ inches wide. At last I hold a journal a little way from the ledge upon which the kitten steps and is lowered by me to the floor.

Sucks mother now practically only at night, though up to the present it has eaten no solid food.

39th day.—When lapping milk the old difficulty is noticed to-day and occasionally still.

Climbs on my lap, and thence upon the table by which I sit, making use of the table-cloth as a means of ascent.

Gets from an ordinary chair to the table, and comes upon my lap by a little leap—the distance being about 7 inches—first crying, however, as if aware of the danger and difficulty.

If at any time it is out of the room a little way, and hears footsteps, etc., *runs* back.

Now goes from table to any chair that may be near it, and *thence to the floor* by a sort of scrambling descent.

On dangling a string before it, the kitten at first *hisses*, and then in a moment *plays* with it, catching it with her paws again and again.

Goes to pan spontaneously. Always cries before answering nature's calls, and paws the sand invariably afterwards, and sometimes before.

40th day.—Weight 1½ lbs. Growing well.

Tries to get on a low chair about a foot from the floor, on which there is a stool that almost covers the top. In this it fails. The kitten then goes to the bookshelf near, and tries to reach the chair. The stretch is too long, and the result is the first and only *fall* I have ever seen her sustain.

When sitting on my knee its mother calls for her. The kitten crouches as if *to hide.* Saw something akin to this lately, when she was sitting in a big chair, and the mother was seeking her.

Runs into the hall, thence to the bedroom, and when scared out *does not* at once return to the study.

To-day *jumps* from the table to the chair, 16 inches below the table, and about 12 inches out from it.

Climbs up the cane back of a rocking-chair, and returns by the same method easily.

Hearing seems to have reached its maximum acuteness, as a very slight sound arouses her, even when drowsing.

41st day.—On entering the room this morning the kitten walks towards me. As I stand, it mounts on hind-legs and puts its fore-paws against my leg.

Is on the table when I am writing, and manifests much interest in the movement of my pen.

After using the pan to-day, covers up the *actual spot* wet for the first time, though always paws somewhere.

When I enter the room in the afternoon, the kitten gets down from a rocking-chair and comes to meet me.

When sitting on my knee, 3 feet from the table, it manifests a desire to get on the table by looking towards it and crying.

42nd day.—Has been out in a dark corner of the hall leading to the study, and in a bedroom also not far off.

Climbs up my leg, and thence to table, three times in succession, though put down each time.

Sits on the table watching the movements of my pen, or playing with various objects it finds about.

Again visits the waste-paper basket, and plays with some crumpled papers in the bottom.

Smells keenly at my fingers that had handled a *recently killed mouse.*

When the mouse is presented to it, the kitten smells it, licks it, and then bites it. When I attempt to draw the mouse away, the kitten holds on with its teeth and claws, and growls slightly.

Jumps down from a height of 1½ feet to-day without any fall.

43rd day.— Kitten sleeps much to-day. When awake wishes to be either in my wife's lap or on the table.

Offer it some meat. Smells and licks it, but makes no attempt to eat it. Licks its lips long after the meat is put to its mouth, as if to get rid of the last trace of the meat. When it is brought near it again, makes off.

Is very playful this evening. Interferes so much with the pen as one writes, it is impossible to do so while it is near.

44th day.—Plays vigorously, but not long at one time. Sleeps a great deal.

In the evening finds a new hiding-place, and fails to make any sign when called.

Is put in pan, but will not remain in. Micturates *in a corner,* and is given a mild punishment, which is well borne.

In the evening visits the pan, and after using it takes great pains to paw up the sand into a heap.

Catches flies. Plays with one of them after it is dead.

Runs into the *most distant room* on the flat on which the study is, *i.e.* it goes further from home, so to speak, than ever before.

45th day.—Races about much to-day, under chairs, out of the room, etc.

Of its own accord visits the pan, and defæcates twice, taking pains to cover up what is passed.

46th day.—Eats for the first time to-day small piece of cooked meat and potato; licks the plate on which it is, and seems to like the food very much.

Acts with a dead mouse as before.

Plays with its mother so violently that the latter seizes it and holds it down.

47th day.—A piece of catnip that affects mature cats so peculiarly, produces no such effects on the kitten, which seems rather to dislike the smell.

48th day.—Full of an apparently reckless activity. Climbs up and down chairs, etc., with great vim and rapidity, but never falls. When tired, sleeps, and leeps long.

Meat is offered to-day. Smells at it, but does not eat till a small piece is placed in its mouth, when it *eats* with apparent relish all placed before it.

When on the table a spool falls to the floor. In an instant the kitten leaps from the table to my wife's lap, and thence to the floor. In its play runs behind the book-shelf, and rushes out at its mother from behind the curtain again and again.

49th day.—Catches flies on the window; takes no notice of dead flies.

When engaged in this sport it upsets a vessel containing a little milk, and is so startled by this that it rushes down at once with a peculiar expression on its face.

50th day.—When we are at dinner to-day the kitten runs into the dining-room on the flat below its home.

Since this long wandering from home it is difficult to keep it in the study, as it wishes to be here, there, and everywhere, maintaining ceaseless activity when not asleep.

Jumps from the table to the floor by one *clear leap.*

51st day.—The kitten climbs from my chair up my back, and rests on my shoulder.

As its mother walks upstairs, the kitten crouches back from the top, as if to spring on her by surprise.

52nd day.—Again climbs up my back, and remains there while I walk downstairs.

On coming home late I go into the dining-room, into which the kitten also shortly walks.

53rd day.—I can notice a change in the shape of the kitten's head. It is more elongated, etc.

Its eyes are also changing colour—getting lighter.

54th day.—Is very playful all day.

At night,-late, when I go into the dining-room, the kitten follows, and, when there, climbs half way up a wire-netting door separating the kitchen from the dining-room.

On stroking the kitten to-day it is noticed to *purr* for the first time.

55th day.—On stroking the kitten it purrs. Soon after, it sits on the table and purrs of its own accord, still louder.

56th day.—Manifests an unusual desire to sit in my lap, and on the cushion of the chair between my legs.

57th day.—The kitten *eyes* very closely, from the top of a rocking-chair, a canary *bird* in a cage at a height of 2 or 3 feet, and at about 4 feet distant.

It also *crouches* when it sees flies moving on the floor, etc.

A small dachshund dog is brought into the study to-day. The kitten puts up its back, growls, etc. The little dog and the kitten are then taken on my wife's lap. The kitten holds off, but looks interested, and finally approaches and smells at the dog.

Spontaneous purring occurs this evening.

The kitten "washes its face," first with one paw, then the other, in an elaborate way not seen before.

58th day.—It is beginning to be difficult to keep the

kitten out of the dining-room, and to-day it tries hard to get in by the wire-gauze door.

59th day.—The kitten seems to be *uneasy* when quite *alone.*

Came *upstairs* to-day to use the pan.

There is a fire in the grate in the study for the first time this autumn. The kitten touches the poker, which is hot. It hisses, etc., but soon after it touches it again and again, in its usual persistent way.

60th day.—After tea I lie down on a sofa in a room adjoining the dining-room, whereupon the kitten climbs upon the sofa, and walks from my legs up to my chest, where it remains purring.

Now generally comes when called "Puss! puss!"

Present physical condition:

Colour not very greatly changed. Is still a light tabby.

The shape of the *head* is now much more like its mother's.

The *eyes* seem to be getting lighter.

Weight, $2\frac{1}{4}$ lbs.

Height, 7 inches.

During the past two weeks it shows much waywardness. When an attempt is made to thwart the kitten in any way it bites, scratches, etc., in a half-earnest, half-playful way.

63rd day.—Slips, by some mishap, in part into a water pitcher, which is followed by hissing and a great fuss generally.

65th day.—For the first time is allowed outside on the front steps. No longer remains as much in the study.

67th day.—Kitten to-day micturates in the *corner* of the study, notwithstanding that it had been solicited to use the pan. Is punished with a small strap.

70th day.—For the past two days the kitten is much inclined to *repeat* the offence of the 67th day.

Very active and very *mischievous*.

Seems to take a delight in interfering with my reading, writing, or whatever I may be engaged in.

Late in the evening I call, "Puss! puss!" but no puss comes. After looking here and there it is discovered beneath the table, beside the waste-paper basket, trying to *hide* from me. When I go to one side of the basket the kitten betakes itself to the other.

77th day.—Is plainly growing fast, especially in height.

The kitten, up to the present, has been out of doors in all about half a dozen times. On one occasion cried to get out.

The kitten still cries when it feels any want, or when it cannot accomplish its object, or is disappointed, *e.g.* it will *cry* if it goes to the window and cannot find flies to catch.

A Skye terrier is brought into the study, also the dachshund that was in before. Seems more inclined to attack the Skye than the other dog.

Is more especially fond of sitting in my wife's lap, and begins to show *attachment* to her above all the other members of the family.

Now likes to lie before the fire. This especially shown yesterday, a wet and cold day.

Now shows fondness for *fish*.

Now often, when put in the pan, will scrape up the sand, yet keep looking about awaiting a chance to escape.

The kitten now takes its sleep mostly at night. It has a much more intelligent, older, more matured look than a couple of weeks ago.

81st day.—When calling for it at night in the dining-room it again plays at hide-and-seek.

84th day.—Physical growth evident. The kitten's *eyes* are now the colour of its mother's.

Shows increased fondness for sitting in my wife's lap, near the fire, etc.

Less trouble for some days in getting the kitten to use the pan.

The dachshund puppy is brought into the study. The kitten manifests less excitement; evidently *remembers* him.

The kitten can now, and for at least three weeks past, watch a *long series* of events *intently*.

87*th day*.—In carriage of body, tail, etc., very like a mature cat.

88*th day*.—The kitten is offered some bread and milk, which it sniffs at, but evidently does not want. After thus smelling at the food it *scratches around* it, as if to cover it up.

91*st day*.—Now, when allowed out the front door, the kitten runs across the road to a vacant lot opposite without special fear. When let in, follows one up close, as I notice a mature cat often does.

The way in which the kitten makes a choice as to a difficult leap to-day, after much evident observation, apparently careful weighing of difficulties, etc., before the decision and choice between several possible ways are finally made, is a fruitful study.

The same judgment is shown in its behaviour towards a *parrot* kept in the same room. The parrot has always proved more than equal to any of the kitten's advances, and this the latter seems fully to realise.

98*th day*.—Growth in height noticeable, though I see the kitten daily so much.

The canary's cage is set in a new position near the parrot. The kitten makes a spring on the cage and falls back. It is punished for this, and takes it well.

Later, it moves up towards the parrot, but the latter is ready to nip its paw, etc., so the kitten withdraws, showing some dread of this quiet but watchful creature.

The kitten still shows the mischievous tendency. It likes to knock down spools from the table, and especially to take pins out of the cushion.

Will sometimes *cry* for my wife to sit down so that it may lie in her lap.

101*st day*.—Was outside this morning when my dogs were being taken out for exercise. Is taken by surprise, and rapidly climbs a tree just at hand, to the height of 30 feet. Does not seem to know how to descend, or fears to do so. A long pole is hoisted up to it, and, after being pushed a little with the pole, it takes the hint and gets on the pole, and so is lowered down.

105*th day*.—The most marked feature in the kitten's conduct for a week past has been its sociability, or its pleasure in being near those human beings with which it has been most associated. Seems to look now to my wife as its best friend.

When lying on the floor near a rocking-chair in motion, the kitten puts its paw *near*, but not under the rocker.

The canary's cage is again near the parrot's, and the kitten seems to find it difficult to refrain from leaping at the canary.

Is now very fond of getting out of doors and roaming about.

Last night, as soon as the lights were put out, the kitten cries, as if wanting to get out of the study, wishing to go downstairs; does so, and at once goes to the front door.

Was taken up again struggling vigorously. When placed in the pan it at once makes use of it.

The kitten seems now to show a great increase in her liking for us all.

To-day, by a series of climbings, gets to my shoulders, then puts up its paws on my head, and purrs in a most decided way, suggestive of *pleasure* and *affection*.

107*th day*.—Is trying to get into a flower-pot with earth in it, apparently with the object of using it as the pan; and about two weeks ago was observed to *scratch* at sand in a vessel set near the fire to dry.

110*th day*.—Kitten sits in bedroom window looking out, and evidently enjoying the sunshine and view.

112*th day*.—Was again found up in a tree very near the house, though the cause is unknown.

The dachshund was in the study to-day. The kitten evidently *remembered* the dog, and they at length put nose to nose.

When the kitten looks down on the dogs from the upper veranda, it shows some fear, begotten of its tree experience, I think, rather than from instinct alone.

Later, the kitten gets sufficiently familiar with the dachshund to be inclined to play with its tail and take other liberties.

After being fed of late the kitten quiets down, seems pleased, perhaps grateful, purrs, etc.

118*th day*.—In order to test its behaviour, a *living mouse*, tied by the tail or hind leg, is confined in a pasteboard box. The kitten smells at the box eagerly, and follows up the box as it is carried away. When the mouse is released the kitten seizes it with a growl. It does not at once kill the creature, but plays with it. When I go near the mouse it is seized with accompanying growling. After thus playing with the mouse for from fifteen to twenty minutes, the kitten eats it completely.

119*th day*.—The dachshund is brought into the study to-day. Is less afraid, and inclines to be more aggressive with the kitten.

The kitten is inclined to play with the dog, but very little seems to bring forth a hiss.

After the kitten has been out in the cold (latter part of November), it expressed great pleasure on its return to comfort, as manifested by purring, rubbing itself against one's legs, etc. .

When any member of the family goes downstairs late in the evening, the kitten will also go (food).

127th day.—The dachshund brought into the study to-day. Both lie on my wife's lap. By degrees the kitten takes more and more liberties, such as biting its ears and neck playfully.

Both go down to the dining-room when dinner is ready. A plate with meat and potatoes on it is set down before them. The kitten snatches a piece of meat, but does not venture to take any more food from the plate.

Later, a Skye terrier is brought in, but the kitten does not make free with it.

Now follows my wife about much like a dog, going after her even out of doors.

When standing near the table the kitten jumps from it upon her shoulder.

129th day.—Flies are now very rare, and whenever the kitten sees one it makes for it.

Sits in the window watching snowflakes fall, and tries to catch them through the window-panes.

To-day, a member of the family the kitten seldom sees, goes into the cellar to get some meat for the cat, and as he walks down the steps he calls, "Puss! puss!" The kitten seems puzzled, and runs about looking up at my wife in the kitchen and crying, and this is repeated, and every time with the same result. My wife goes to the cellar door, induces the kitten to go down a step or two, when the kitten, observing the one who is calling it near the meat, runs down.

135th day.—The kitten is out early in the evening in front of the house, and is after a while sought for but cannot be found, having, in all probability, been picked up by some one passing.

Remarks on the Diary.

Since I propose to make a comparison between the development of the dog and the cat in another paper, by which more instructive results may, it is hoped, be reached, the remarks that follow are to be considered rather in the light of suggestions as to some of the ways in which the diary may be utilised than as a commentary at all complete.

It will be seen from the records that the evidence for taste and smell before the 3rd day is not very convincing; that the evidence for a temperature sense, the feeling of pain, ordinary reflex action from tactile sensibility, etc., is more conclusive.

By the 6th day a great *advance* is recorded in regard to all these qualities.

Fatigue is still readily experienced.

Sometimes, as *e.g.* on the 9th day, an apparently decided advance is shown in a single day.

The experience of fear or surprise seems to be experienced first about the 9th day through the sense of hearing, if we except that uneasiness and crying that results almost from the first when the animal is removed from its usual surroundings.

The spontaneous movements of the ears, and more doubtful movement of the tail, on the 10th day, are worthy of note.

Attention is called to the advances in vision as noted on the 11th, 13th, 14th, 18th, 22nd, 25th, 26th, etc., days.

On the 8th day the eyes began to open, but hearing was then established.

On the 11th day winking is caused reflexly by the

movement of an object before the eyes at the distance of a ½ inch.

On the 13th day the pupils respond to light.

On the 14th day a small object is followed by the eyes.

The 16th day was memorable, as on that day was noticed the first licking of the paw, the first act of scratching, and the first play.

On the 18th day the nose, uncoloured at birth, began to pigment.

The 21st day furnishes evidences of recognition by the kitten of its box as its home, for, after being out, it returns to it and climbs in.

Attention is directed to the various stages of development of muscular co-ordination, as illustrated by the kitten's toilet-making, play, getting in and out of its box, and from one object to another in the room, and which can be followed from the somewhat full records of facts bearing on this subject. The records of certain days are clear on this point, *e.g.* 16th, 18th, 21st, 23rd, 28th, 29th, 30th, 32nd, 33rd, 37th, 51st.

The behaviour of the kitten towards the same dog at different times, and as compared with the second dog, seems to me to indicate an interesting struggle between instinct and other qualities, and shows how the result, in any one case, depends on past experience, the stage of development, and much more that each reader will put into the case, according to his own views of physiology and psychology.*

---

* Mr T. Mann Jones' remarks on this subject will be read with interest: "You deal with the notion of hereditary (psychical) memory and aversions more gently than I should. As early as a boy I had come to the conclusion, from experiments, that if I allowed hereditary memory of dogs to kittens, I must also allow they had hereditary memory of a chemical laboratory; and later, my experiments on fowls convinced me if I admitted memory (psychic) of the hawk's cry, I must also admit similar memory of the report of a gun and the

Similar remarks may be made as regards the behaviour of the kitten towards the sand pan. This little history illustrates, it seems to me, some of the fundamental laws of all training and education, whether applied to the human being or the lower animals.*

The case is simpler as regards the latter, but not wholly different, and observations of the kind made in this case impress me more than ever with the importance of attempting to give the fullest possible record of every feature in the psychic development and the physical development of those animals by which we are surrounded.

The history of the kitten's whole bearing towards the book-shelves has been to me a most instructive one. I have never witnessed such perseverance in the accomplishment of an object in any young animal

---

sound of watchman's rattle and 'penny squeakers.' It was a *reductio ad absurdum*.

"In 1872 and 1873 I reared two small broods of chicks by the aid of a large tom-cat, Pedro, who acted as *foster-parent*, and in whose fur they nestled when cold, and slept at night. When half-grown, or less, they would run towards any cat, and were attracted by the voice even if they did not see the animal, and would have been killed but for the watchful aid of Pedro; yet, they avoided all my hens, and took refuge hastily with Pedro if a 'clucking' or broody hen called them, or a cock crowed near them!"

* Mr T. Mann Jones writes me as follows: "I was greatly pleased to see you had noted the rise of the habit of cleanliness in the kitten. I have given considerable attention to this in children, dogs, and cats. I have been often surprised that no naturalist whose works I have read had given the subject the attention it probably deserves. From personal observation, and such information as I could get from mothers and nurses, I have come to the provisional conclusion that early self-control as to the sphincters, whether in man or animal, is a pretty fair indication that the adult will possess more than average self-control. What a difference in setting up 'cleanly' habits corresponds with in later life of self-indulgence and ready acceptance of every suggestion to mere instinctive action, *i.e.* we have a weak-nerved animal more or less, though often a very obstinate one.

"May I suggest a probably fresh field of observation in animal psychology connected with this? The efforts of the long-haired dogs to avoid uncleanliness when suffering from diarrhœa. Punch avails himself of a wire fence, using the second wire as a skirt holder, by *carefully* adjusting his long fur on it, and the lower wire as a seat! ... The plan answered. Great gratification followed."

—not excepting the child. It seemed that the greater the obstacles, the greater the efforts the kitten put forth to overcome them—behaviour that we usually consider especially human, and ever an evidence of unusual strength of character.

That this kitten was not an ordinary one in many respects I am quite prepared to believe, but still the animal was a cat, and a cat only, and that such "character" should have been shown was a surprise to one who has been a long and close observer of animals.

I have seen something akin to this in that remarkable bird the parrot, but not in a parrot so young as this kitten.

One of the remarkable features in the whole group of the *felidæ* was illustrated in this kitten, viz. the slowness with which they learn to eat and drink, and the length of time before difficulties are fully overcome.

A comparison of the kitten's behaviour towards the parrot and the canary furnishes food for reflection, and in this, as in all such cases, all narrow explanations prove inadequate; and while the laws of association etc. may explain much, they do not seem to me to explain all in the case of the lower animals, any more than in the case of the child or the man.

Conclusions.

While there can be no doubt that cats are born deaf and blind, the question of smell and taste is more difficult to settle. Up to the third day, and even then, there is no clear evidence of smell and taste, though, on the whole, it would appear that the facts in favour of the existence of smell are more certain than in the case of taste.

There is evidence, on the 3rd day, of reflex action,

brought about through stimulation of the skin; of the existence of the ability to distinguish hot and cold objects, and of capacity to suffer pain; though, as regards all these, the latent period is relatively very long. At birth, and up to the 3rd day, though it is likely that all these powers exist in the animal, the evidence is feeble.

Motor power is restricted to such an extent that the animal can crawl only and but slowly.

Tactile sensibility, the sense of pain, and the temperature sense reach their highest degree of development earlier than hearing and vision.

Hearing is established before seeing, and reaches its greatest perfection sooner than vision.

In the progress of all the senses to full development, the course, while marked by definite steps, is often so rapid that distinct advances may sometimes be noted in a single day.

Not only in the development of the senses, but in all other phases of progress has this been clearly evident in the case of the kitten under observation.

Apart from the senses, etc., there seems to be a definite order in which all the features of feline nature appear, *e.g.* purring, crouching, stalking, etc.

Certain physical changes are correlated in time with certain psychic developments, the significance of which is in some cases clear, in others obscure.

III.—THE MONGREL DOG.

The Mongrel and the Pure-Bred Dog Compared.

IN my first paper on the psychic development, etc. of the Dog, the observations and conclusions applied to pure-bred animals only, though two widely different breeds were compared.

It seemed to me desirable for many reasons that the mongrel should also be studied, accordingly this paper will be devoted to that purpose, and to a comparison between the mongrel and the pure-bred dog, in a manner to harmonise with my general plan of treatment of the subject of psychogenesis in the lower animals.

Diary.

The puppies under observation belonged to a litter of nine, of which seven were preserved. The dam was a strong, healthy black animal, and the sire was unknown. The dam and puppies were kept under similar conditions to those described in my first paper on the Dog, *i.e.* in a pen on the floor of a separate compartment of the kennel, on which at first there was abundance of good straw, and at a later date sawdust. The puppies were born in December, and artificial heat was maintained in the kennel constantly, so that the animals were always comfortable. The dam was well fed, and otherwise properly cared for throughout, and from beginning to end was perfectly well, and able to nurse her puppies in an entirely satisfactory manner.

1*st day.*—Vigour of puppies shown by a strong *voice*, somewhat between a growl and a bark.

They respond to a *prick* or punch and to a *hot* body, but not so quickly. Oil of wormseed placed near the nose causes a sniffing; pure carbolic acid causes the face to be distorted; blistering fluid leads to protrusion of the tongue; and tincture of iodine to sniffing.

When soup is similarly placed before the nose there is *no* evidence of *smell*.

Solutions of salt and of sugar, also cow's milk, are presented at different times. Certain conclusions cannot be drawn. The brush on which the solution of sugar is presented, is sucked vigorously, however.

The puppies will *not crawl off* a table, etc.

2nd day.—The irritating substances used on the first day produce more marked effects.

No positive evidence is forthcoming as regards milk, but sugar seems to be agreeable.

When salt is put in the mouth it is rejected doubtfully, but aloes most *decidedly*. Both are used in solution.

5th day.—Irritants affect the puppies through the nose at a greater distance.

It remains doubtful whether *meat* is smelt or not, but milk and a solution of aloes seem to be, for the puppies cease to make their characteristic sounds when these substances are placed before the nose.

They do not yet seem to show any shortening of the latent period of reflex action on pinching.

9th day.—None of them show any signs of opening of the eyes.

10th day.—When blistering fluid in a bottle is brought near, active movements result; the head is lifted up, the tongue put out, and sneezing, etc., follows.

Now, when a solution of salt of the same strength as that used before is presented, it is *rejected*, and there is frothing at the mouth, etc. Much disgust is also shown towards aloes; with milk the result is not pronounced, but with sugar there is undoubted enjoyment.

Eyes begin to open.

19th day.—On clapping my hands, and uttering "Hist!" vigorously, the *ears* are moved, indicating the ability to hear.

20th day.—Evidence of hearing, as above, at the distance of 4 feet.

A piece of meat, put within a $\frac{1}{2}$-inch of the nose of one of the puppies, causes it to move towards the object and lick the lips. The same follows when cheese is used. When the same experiments are tried with sleeping puppies they also lick the lips.

Stamping with moderate force on the floor on the kennel (concussion), within 2 feet of the puppies, rouses them far more effectively than a sound of considerable loudness.

One of the puppies is brought into my study. It moves about, crawling rather than walking, with its *tail* carried or held, much as in a turtle, the *hind legs* being much spread out.

No great uneasiness manifested.

The puppy is a picture of healthy vegetative existence.

The *incisor teeth* are appearing.

As an evidence that the puppy is influenced by the change in the *environment*, it may be mentioned that the cheese and the meat do not produce now the effects they did when the puppy was in its own pen.

Under favourable conditions a mere trace of salt causes decided signs of disgust.

When blistering fluid is held within 2 to 3 inches of the nose the puppy acts as if something unpleasant were in its mouth.

The puppy winks occasionally, apparently spontaneously, but *winking* cannot be induced *reflexly* until the moving finger is approached so close to the eye as to almost touch the eyelashes.

When the finger and thumb are moved, as in snapping the fingers, but without causing any noise, the puppy under observation *turns* its *eyes* in the direction of the object, so that I have no doubt that a small object, as a finger, is distinguished at 3 to 4 inches, and a larger one at about 2 feet. When the fingers are snapped in the ordinary way, the head, as well as the eyes, are turned towards the exciting object, showing that vision and hearing are both stimulated effectively and together.

Motor activity is still poorly developed.

The coat, straight before, begins to show waviness.

24th day.—To-day a puppy follows a fish-bone by smell, and attempts to bite at it.

25th day.—The appearance of the eyes show an advance over a few days ago.

When the dam is brought within 6 to 8 feet of the puppies they seem to become aware of her presence by *smell first* of all, but when within a few feet of them they follow her with the eyes.

There is a decided improvement in locomotion.

The puppies now have all the upper incisors, the canines above and below, and two molars above and two below.

Hearing is now very acute. A very feeble "Hist!" when within a couple of feet, causes movement of the ears, turning of the head, and an expression of the face that indicates clearly that the puppy hears.

A "Hist!" of moderate intensity is heard readily at 5 feet.

There is a tendency now, when any object is placed before the puppies, to *growl or bark* at it in a meaningless way.

31st day.—To-day a slight sound caused a starting. The puppy (henceforth the only one) seems to have reached that period when hearing, as a mere sensation, is perfect, but the interpretation of sounds very imperfect, hence he is startled by many sounds that later produce no such effect. When a bone is placed within 4 inches of him he sniffs at it, and gets up to secure it.

He makes the first attempts at *gnawing* a bone, using his paws to steady himself, and once lifts a paw as if to place it on the bone, but does not actually do so.

The puppy now evidently sees small objects at some distance quite well.

He now stands quite erect on his legs, and to-day climbs over the rungs of a small rocking-chair.

When in the house to-day he manifests *shyness* or appreciation of the strangeness of his surroundings.

When his head is *stroked* he acts somewhat as a cat.

Has begun a sort of *playing* with his dam.

The *social* tendency is clearly evidenced.

38th day.—Has grown much. There is a great change in his attitude, manner of walking, etc. The puppy can now *run* fairly well.

He has wonderful control of his *tail*, moving it when he approaches vigorously, as an old dog.

For a week past his *social* tendencies towards human beings have been very marked.

He is now provided with, practically, his full complement of *teeth*.

His *eyes* and general *facial expression* have also greatly changed.

He is still readily startled by sounds.

He as yet reposes a good deal, which perhaps accounts in no small degree for his perfect health and excellent state of nutrition.

He now sees small objects at a distance of 8 to 10 feet at least.

He is ready to *eat* almost anything given him.

He can now *bite* accurately at any part of his body.

39th day.—His *weaning* is begun regularly to-day.

42nd day.—The following experiments were tried to-day : When a finger was put in his mouth he sucked it, but he would take no notice of a stick presented in the same way. Bread he tackled at once, using his back teeth to crush it. Meat he devoured eagerly. He had received no nourishment of any kind for one hour previous to these experiments.

45th day.—Puppy is brought into my study. Shows little or no shyness after a few moments. He is inclined to move about and use his nose in an *investigating* way.

He can now run fast and well, his power over his *hind legs* being good.

He hears perfectly even faint sounds.

On being held before a piano, when it is played upon, he manifests no special effect.

46*th day.*—The puppy gets up to a water-pan over a foot high, and, resting his front paws on it, laps water, the whole proceeding reminding one of a mature dog. This was first done, however, three or four days ago.

A small Skye terrier, and afterwards a collie, are allowed to enter the puppy's compartment, but he does not seem inclined to notice them much.

Lies with his head on the sill of the door of his compartment.

47*th day.*—After I play a little while with the puppy, and then retire, he barks impatiently to get out.

49*th day.*—The puppy is playing in the yard. Scratches at the snow like an old dog. After defæcation he draws himself along in the snow.

He pulls at the withered branch of a vine growing against a fence.

He ascends a snow bank in the yard, wagging his tail.

He seems very much at home with the older dogs, and adapts well to his surroundings.

52*nd day.*—The last remark applies with much greater force now.

He is very free in all his movements. Carries his *tail* up, and wags it vigorously.

Considerable change is observable in the shape of his *muzzle* within a few days.

On being given a *bone* he does not commence on it at once, but carries it to his bed instead of gnawing it on the sawdust that covers the floor of his pen.

53*rd day.*—To-day the puppy is trying hard to get out of his compartment, pushing his head through between the iron-wire door and the sill, and using *his paws* to enlarge the opening—in fact, he behaves in this very much like a mature dog.

56*th day*.—When he is out of his compartment (the door open) and he is told to go in, and one stamps with the foot, he retreats to his box, but soon wants to come out again.

63*rd day*.—He now mingles with all the dogs of the kennel, large and small, seems to enjoy the life, and manages to adapt admirably to his entire environment.

69*th day*.—Prior to defæcation moves about, smells, turns round and round, etc., just like an old dog.

Shows sexual feeling, if one may judge by his actions towards a mature bitch.

87*th day*.—Since the last record, his progress in adaptation and in general development has been steady. As the chief stages in development had been now overtaken, he was removed to my laboratory for experiment on the brain.

Remarks on the Diary.

Observations on the temperature sense, tactile sensibility, and the sense of pain were not made as early as desirable in the pure-bred puppies, but this omission was made good in the present case, as the diary shows, and there can be no doubt that these senses all exist from the moment of birth. The evidence that true *smell*, as distinguished from the mere sense of irritation of the nose by pungent vapours, is less conclusive, while that for taste is more doubtful still.

The increasing development of tactile sensibility, and especially the readiness of the nervous system to respond to stimuli acting on the nasal mucous membrane, is well illustrated by the observations recorded at different times in the early life period of the puppies.

Both smell and taste are very feebly developed before the eyes are open, but about the 10th day there was clear evidence of both enjoyment and disgust through taste, at all events.

The general advancement of the animal is well shown in its behaviour towards blistering fluid on the 10th day.

By the 20th day smell had become a powerful moving force in the animal, as it always will continue to be.

The manner in which a sleeping dog, young or old, is affected by the presence of food with a pronounced smell, is very impressive to one who witnesses it.

At so early a period as the 24th day this puppy's sense of smell was so well developed, it will be noticed, that he followed a bone by its aid quite well. At this period he used his eyes, as well as his nose, to guide him.

At this date also hearing was good—in fact, by the 25th day the dog had reached a period of fair development of all his senses, and with considerable motor power, which, of course, also implies a corresponding development of the muscular sense. Hearing seems to be the most rapidly developed of all the senses, *i.e.* the period from its first beginnings to its greatest acuteness is relatively short.

The pleasure of the puppy on having its head stroked, on the 31st day, is noteworthy.

The Mongrel Dog and the Pure-Bred Dog Compared.

In my first paper on the dog, certain differences were noted between two varieties of pure-bred dogs—the one large and the other small, viz. between the St Bernard and the Bedlington terrier. I now propose to compare the pure-bred and the mongrel, chiefly on the basis of my records in the earlier and in this paper.

The mongrel showed more vigour at birth, and shortly after, as evidenced to me at a distance, by the voice. He also seemed to be somewhat less influenced by cold, though more persistent, or at least more successful, in getting all the heat possible from the dam.

Although this is not specially noted in the diary, the mongrel did not show fatigue quite as soon as the pure-bred dog.

The eyes began to open earlier than in either the St Bernard or the Bedlington. I could not get certain evidence of smell in the case of the St Bernard before the 12th or 13th day, but in the mongrel there is some evidence of smell on the 10th day, at latest. The Bedlington smelt meat on the 10th day. I am not prepared to state that dogs do not smell at all before the 10th day, or even the 5th day, for which there was some evidence, or even on the first day, but I am of opinion that before, at least a few days after birth, smell is so feebly developed, if at all, that one could not *demonstrate* its existence to an unbiassed observer. I refer now, not to the effect of irritating volatile liquids, but to the ability to smell food.

The mongrel showed signs of disgust with aloes, and a positive liking for sugar, as early as the 2nd day. I could not establish this for the pure-bred dogs for some days later.

It is noteworthy that, on the 3rd day, the mongrel sucks one's finger more readily than a piece of meat.

Though dogs are carnivorous animals before a certain period in their development, they are no more excited by meat than any object whatever, showing, in the clearest way, that there is an order in psychic as in physical development.

Owing to absence from home when the mongrels were passing from the 11th to 18th day, I am unable to make any accurate comparison in this case as to the beginnings of vision and hearing.

In both the mongrel and the pure-bred dogs, the winking reflex is very slowly developed, and long remains feeble, *i.e.* is excited with difficulty.

Certain considerations must be borne in mind in

attempting to compare the mongrel puppy and the St Bernards. The latter, I consider an unusually active litter, while the mongrel, for a considerable period, seemed to me more than usually vegetative. Moreover, while there were always at least four St Bernards together, this mongrel was the only one of this litter after about the 20th day.

One of the features of development greatly impressed on my mind by these comparisons, not to mention many other similar ones, was the influence of one on another in all the lines of development. This was shown both negatively and positively in the case of the mongrel. After he began to mingle with the older dogs his progress was marvellous. He seemed in a few days to overtake himself, so to speak, and his advancement was literally by leaps and bounds. It is very difficult to give an adequate idea of this feature of the mongrel's history in a diary, but I wish to note it specially, because it seems to me to show that, while education, in the wider signification of that word, may in a sense account for development, it is equally true that the real nature of any animal will, in the main, assert itself sooner or later, however unfavourable the early environment. In other words, heredity is, was, and ever will be, stronger than environment.

One may safely say that in all kinds of dogs the perception of light and shadows precedes the seeing of objects, and that the latter is comparatively slowly developed.

The mongrel seems to have been very slow in developing the play instinct, which I attribute largely to his being the sole puppy from an early period, and therefore seeing no other dog but his dam.

In both the mongrel and the pure-bred puppies hearing progresses rapidly to perfection of sensation. Within about ten days the maximum of acuteness is reached, and the puppies are then very readily startled by noises, *i.e.* they are reflexly greatly affected through the ear, much more so than through the eye. This fact has been very strongly impressed on my own mind again and again.

As regards voluntary movements, there were differences to be noted. The mongrel seemed the sooner to gain control over the hind limbs. The same may be said of tail movements. Though one could not trace so general a development in the mongrel as in the St Bernards by a certain period, the former had the greater power over his tail, both as regards variety and vigour of movement, especially the latter.

There was a comparatively sudden development in this mongrel which I can scarcely think is common, but, in regard to this point, I must speak cautiously till further exact observations are made, as I do not wish to make statements of mere general impressions, with no definite basis of observation.

The movement of the ears especially, though others also fall into this class, following on noises, are purely reflex, and constitute one of the most delicate and early evidences of hearing, but, as in the case of the pure-bred puppies, concussions are earlier and more readily effective than sounds.

While both the pure-bred dogs and the mongrel recognise a change in surroundings, and are affected by it, herein lies one of the great differences between them alike in puppyhood and in after life.

One may compare the man "well born," and surrounded by conditions that tend to form the gentleman in the completest sense of the term, with the pure-bred dog, just as the mongrel represents the peasant, whose ancestors and whose surroundings alike are against the highest attainments.

The mongrel lacks all that refinement, modesty,

perhaps I may say, at all events that delicate appreciation of his own place and that of his fellow dogs and human beings, which constitute such conspicuous features in the psychic whole of the pure-bred dogs. The latter show towards each other, in a kennel with many together, when their owner encourages and gives a chance for their exercise, kindness, justice, and many qualities utterly foreign to the nature of the majority of mongrels. I do not now use the term mongrel in the sense of a cross between two pure-bred dogs, but in the more popular acceptation of a dog bred from parents that were mongrels, and perhaps the remoter ancestors quite unknown.

It is interesting to enquire whether these features of the psychic organisation are shown in the young puppy, and if so, when.

Almost from the first the mongrel puppy shows an ability to scramble for himself in "this rough world" not manifested by the pure-bred dog. His very voice on the first day of his existence may, and in this case did, suggest this, but in nothing was it shown so much as in the successful manner in which he held his own among the dogs of the kennel, large and small. This was all shown before the puppy was two months old. His confidence in himself, his power to adapt to unfavourable surroundings, was as advanced at this age as the St Bernards at four months. He reminded me of a forward boy, lacking in all true modesty and due appreciation of what was due to his seniors. Yet this mongrel, by virtue of this very psychic condition, succeeds in his purposes, if one may so express it.

In the litter of St Bernards, the most prominent and precocious one could not compare with this mongrel. In the lower animals, development is so rapid that new features in the psychic character at time seem to reach a comparatively complete development rather suddenly, even when the animal is watched daily. This was especially observable in the mongrel puppy after a certain date, and was, I think, all the more so because his development for a time seemed rather slow, which I attribute in part to his being the sole puppy for most of his life.

And here I would draw special attention to facts of this character. The puppy's weaning was not begun till the 39th day, as his dam was well able to supply him with milk, and his nutrition was excellent, but when this process, generally requiring a good deal of care, and often attended by much derangement of health, was begun, there seemed to be no need of gradation, etc. The puppy was ready at once in every sense to eat all kinds of food fit for a dog.

When he was allowed out in the yard, all voluntary movements seemed to undergo a rapid development, which was not largely due, I think, to special exercise or practice, but to the sudden development of elements of the nerve centres that had been functionally latent.

As before stated, this case of rapid development in different directions has impressed me very forcibly, and seems to be in harmony with a law of nature of pretty wide application.

I do not think this puppy's intelligence was equal to that of the St Bernards at three months, though superficial observation might have led one to conclude the reverse. Forward people often pass with the undiscerning for having an ability they do not possess, because of their confident, showy bearing, and the same remark would apply to this mongrel puppy. The animal is now more than four months old, and I have seen nothing in him to lead me to alter this opinion, though much to confirm what I have endeavoured to make clear, as having impressed me as true of the psychic nature of mongrels, as compared with pure-bred dogs.

The characteristic physical features of the adult certainly appear in mongrels sooner than in all the larger breeds of dogs; in other words, they mature sooner than these breeds, both physically and psychically, though not in all cases more rapidly than the smaller breeds of terriers. While all pure-bred dogs must have a definite rate of development, according to the breed, it must be plain that mongrels will vary much more individually according to the nature of the ancestors which have contributed to their highly composite origin in many instances. In the very nature of the case, the pure-bred dog is an inbred animal, while the mongrel is generally the very reverse. He, in fact, approaches far more closely to the wild *Canidæ* in this respect.

From this it would be expected that the physical changes would be of a kind that would appeal to the eye much less—would be, in fact, less readily referred to any type or pattern; and in no small degree is this true of his psychic characteristics, though these things are much more readily observed than made clear by any records.*

---

* I append Mr T. Mann Jones' remarks, which grew out of a discussion of the question of latent or undeveloped power in both human beings and animals, though they will suggest much beyond that, while they constitute one of the most interesting animal histories with which the author is acquainted: "This latent power is apparently called forth in some dogs to a great extent by slight stimulation. My grandfather selected pups of different breeds, among others Newfoundlands, mastiffs, and bloodhounds, as he distrusted the ordinary view of 'instinct' as the cause why certain breeds would attend to sheep, or were sheep-dogs, and he found in each race individuals who made as intelligent sheep-dogs as those who 'were to the manner born.' When I had my opportunity I made observations, and I sought information from farmers and shepherds. They say: 'Any dog of any breed which is powerful enough bodily, makes a good sheep-dog if *he will try to be one.*' 'Special instinct' appears here to be an unnecessary cause. To exemplify by an individual. In

IV.—THE DOG AND THE CAT COMPARED.

ALTHOUGH, in popular estimation, the dog and the cat are considered as opposites in almost every respect, in reality they have much more in common than any two of those animals commonly kept by man, as should be expected from their place in nature.

A comparison will, however, prove profitable, it is believed, and this will be based chiefly on the diaries of the papers on "The Dog" and "The Cat" respectively.

Both the dog and the cat, it is scarcely necessary to

Appendix D. of Herbert Spencer's 'Justice,' there is some account of Judy and her *pup*. Punch is the monkey who was five years old when I wrote some letters to H. S., which he inserted in the 'Justice.' He commented on Punch's actions, calling him 'The Christian Dog.' The pup was suckled almost in the dark, and the mother removed to some seven miles off as soon as he could lap milk, to prevent the rise of imitative habits.

"I could not note that his head differed notably from Jack's (the sire) at two years of age, nor from the mother's, so far as I recollected, but 'strain' of the environment told on him. Somehow he discovered that one of the family could not hear, and took her under his charge, and for eight years he has never let horse, vehicle, or tramp approach her without warning, and if she misunderstands his indication as to direction, he either stops her altogether, or, if there is danger, hustles her, and in some urgent cases, by no means gently, out of the way. But the unaccustomed environment told in another way. In consequence of accidents I have had to walk with a crutch. I came to a slippery place one day, and slightly slipped. The dog shrank as if struck; he had been running finely over such places freely before, but directly I came near another place of the kind the dog assumed the 'danger step,' or 'cautious step,' looked up in my face, and walked with this step which dogs naturally use when approaching an unknown and possibly dangerous object. I was much puzzled one day by his use of this gait, and his 'edging me' out of the direct line I was taking, as there was no apparent reason for it. On standing and looking back, he at once pointed, and on examining the grass I found that a brood of very young geese (without the parent) was hidden in the long grass, and if he had not turned me I must have stepped upon them. The dog's sire was a pure otter dog, mother a mixed breed, with much of the Spaniel in her. It is doubtful if there is any pointer blood, yet the dog is always pointing or indicating by other gesture—not birds, however, but things we should not otherwise observe. A slip among the rocks of those he is attached to, if very definite, calls forth a positive *shriek*, followed by his hastening without regard to falls and tumbles to the one who has slipped."

point out, are born blind and deaf, but the eyes of the cat open sooner than those of the dog, and hearing is also acquired somewhat earlier, but in both the processes of learning to see and to hear are gradual ones.

The pupillary reflex is established sooner in the cat.

So early as the 9th day the kitten studied and previously reported on turned its ears towards the direction of a sound.

There is this difference, too, in the movement of the ears: the kitten, when reacting to a sound, turns the ears or ear reflexly to the side rather, while the dog tends to draw them back.

I have observed nothing in the young dog that corresponds to the quivering movements of the ears in the kitten, seen as early as the 9th day, and which possibly are imperfectly executed voluntary movements, like the trembling of the hand in old people.

There is nothing in the dog that corresponds exactly to the hiss, or when feebler, the opening of the mouth in the kitten when surprised. So far as I know, this is *sui generis* among our domestic mammals, though there are analogies perhaps in birds, as in the hissing of geese or ducks, and the snapping of the beak in pigeons, even when very young, to which abundant reference has been made in my corresponding paper on birds.

So early as the 3rd day the young cat gives evidence of the possession of genuine smell, as shown in its behaviour towards a dog placed near it. At the same time, the sense of the smell is very feeble. Upon the whole, it would seem that taste and smell are both present rather sooner in the cat than in the dog, and in both the beginning is feeble, but they go on to fairly rapid development. However, I have not changed my opinion, as expressed in my first paper on the dog, that the dog, and I will now add the cat, find the nipple of the mother by touch rather than by smell, and that they are drawn towards the belly of the mother by the warmth of the part.

In both the dog and the cat there is a long latent period in the case of reflex movements from a pinch, etc., as compared with such an animal as the rabbit, though there can be no doubt that the tactile sensibility, the capability of feeling pain, and the temperature sense, as well as feeble motor power, hardly worthy the designation (voluntary), exist in the dog and the cat at birth.

I am not prepared either to affirm or deny that taste and smell are present at birth, but if they do exist, I am quite sure they are of the feeblest, of very little use to the animal, and play but a very subordinate part in its life during the blind period.

The kitten is at first, if not always, more sensitive to a touch, has finer tactile sensibility about the mouth than the puppy.

There are the same individual differences as to the exact date of the opening of the eyes, the eruption of the teeth, etc., in the kitten as in the puppy.

The dog and the cat resemble each other in the slowness with which they acquire power over the hind limbs.

Neither the puppy nor the kitten have any appreciable voluntary control over the tail during the blind period, but the dog finally uses the tail much more than the cat in the expression of his emotions. What the dog does with his tail the cat often expresses by purring, which, as I have shown in the paper on the cat, is developed somewhat late—much later than the friendly wagging of the tail in the dog; and as will be seen by a comparison of the notes (diaries) on the dog and on the cat, while there are definite stages in tail carriage for each, these are different altogether for the two animals, and herein we notice a far greater difference than in locomotor activity. The tail movements and carriage are definitely related to the character of the animals, and to those that watch them closely express distinct and varying phases of emotion, etc.

The antipathy of the cat to the dog, while related to a psychic state, based on self-preservation from intruders, is peculiarly marked towards the dog, though whether more so than towards any other similar animal, or, towards, say a large part of the animals that might be found in any menagerie, is one that I have not investigated. I have been very much impressed by the fact, that at an early age the kitten, when suddenly disturbed in any way, reacts much as if a dog had come upon it, though in a less marked manner.

Nevertheless, the behaviour of a kitten, even a few days after its birth, towards even the smell of a dog on the hands, is very suggestive of an instinctive fear or dislike of the dog. At the same time, I have seen a kitten act much the same when an irritant was placed near its nose, or, after it could hear, when it was startled by a noise. This subject is worthy of further study.

Equally striking in the kitten, as in the puppy, is the rapidity with which the creature tires under any sort of stimulus, especially within the first twenty days of life. After a few trials, sometimes after the very first one, the smell of a dog ceases to produce the reaction in the cat during the blind period, and unless one is aware of this, all sorts of erroneous conclusions may be drawn regarding very young animals. This tendency to rapid fatigue indicates, in reality, both why the animals do sleep and must sleep so often. I am quite satisfied that any sort of irritation, whether from within or from without, that will prevent frequent periods of sleep occurring, will disorder the health and even cause death in young animals, and I believe this is one reason why parasites are so injurious to very young animals.

As in the case of the dog, a young kitten, even on the day of its birth, will be slow to crawl off a surface —as a table. These animals have what amounts to a sense of support, the absence of which causes them uneasy sensations. They turn away from the space beyond their support because it does not afford the essential sensation, and as I have remarked in my first paper on the dog, this seems to me as fundamental as anything that is to be found in animal psychology.

In the cat, as in the dog, the winking reflex is slowly developed, and is never so marked as in man. A cat can look at one much more steadily than a dog, and for a longer time, a fact which has its own psychical significance.

The cat knows no shyness or modesty in the sense in which a dog, especially a pure-bred dog, experiences such a feeling.

In one particular the cat is greatly in advance of the dog at the corresponding period, and also finally, viz. in co-ordination of voluntary movements.

Though according to my notes, the kitten did not begin to use the limbs in scratching (quite a complicated movement for a young animal) much before the puppy, if at all, still progress, even in this direction, was much more rapid in the cat. I have taken care to give a very complete account of the movements (actions) of the kitten, so that there might be available a full history of the development of these movements in an animal in which they finally reach extraordinary perfection.

There is no comparison between a puppy's range of co-ordinated movements at three months, and those of a kitten of the same age.

That in the course of evolution the possession of sharp nails has had much to do with this, I feel satisfied.

P

In both dog and cat the cortical centre of the brain for the fore-limb is readily excited by artificial stimulation; but this crude method and general result do not bring out the differences that the animal can by its own will accomplish, and serves, when taken with the facts of the animal's actual life, only to show how very imperfect are our physiological imitations of will-power in these animals.

None of our domestic animals has such power over the fore-limbs as the cat, and this is well established when the animal is even two months old. The development, as my diary shows, is very rapid when once it begins.

And this is closely related to the play of the kitten.

Play is especially instructive. The young animal has an excess of vital energy. Very soon this begins to express itself in imitative actions. I hope my diaries will furnish scope for comparison of the puppy and the kitten as regards play. Herein the animals differ widely, and reflect to perfection their psychic moods.

The crouching, lying-in-wait, the concealment, of which I have made several records for the kitten, are only late and comparatively feebly developed in the dog—all of which is, of course, related to the manner in which the mature animal secures its prey in the wild state.

The *Canidæ* hunt either alone or in packs, and rely on swiftness and concerted action.

The *Felidæ* lie in wait, mostly alone or in pairs, and spring on their prey, so the kitten, when quite young, does not wait for a mouse to appear, but gives its instinct free scope in its attacks on flies, and if these be not forthcoming, it will, out of something, construct imaginary prey for its gratification.

Again, the cat is very slow to develop, as my diary

shows, the social instincts so far as man and other animals are concerned. How seldom a cat seems even to miss its old friends, if indeed they are to it friends. Not that I believe the cat an entirely ungrateful animal. It is very sensitive to good and to bad treatment, but it is not dependent on man either physically or psychically. The cat may, of its own accord, take to the fields and woods to secure an independent existence, and so long as the environment is favourable, it may, it would seem, be utterly oblivious alike of friends and foes.

This independence was shown quite early in the case of my kitten. At the same time, one of the most interesting features in this psychic study has been noting the way in which higher mental states and better qualities prevailed in the end in this kitten, under good treatment. It had finally become social and affectionate, discriminating in favour of the one who had really done the most for its comfort. But of self-denying, purely unselfish devotion to a master, as in the case of the dog, there seems to be little—very little—in the cat. But puss is no flatterer, and her readiness to resent ill-treatment may have had much to do with her not occupying a higher position in man's esteem.

I have, myself, raised a cat from the depth of degradation, so to speak, to self-respect and the respect of others by patient and persevering good treatment, and I am anxious to record the fact, as I believe the cat to be much misunderstood, and its intelligence greatly underrated.

If the term intelligence be employed in a wide sense, and be made to cover the power an animal has to adapt means to ends, in a more or less conscious way, including the adaptation of its own organisation to the environment, then the diary of the cat will furnish an interesting record bearing on this subject. In fact, from this point of view, the cat, during the first three months of its life, is decidedly in advance of the dog.

In the mature cat, instinct in securing prey plays so prominent a part, that we are apt to overlook a great deal in the mental experience of the cat. Her psychic life is withdrawn from us to a greater extent than that of most of our domestic animals. I do not know of a single good history of the complete development of the cat from birth to maturity, so that I regret the more the loss of my kitten before she had reached the age of at least one year.

The diary also shows that the cat has a good memory, though whether equal to, or better than, that of the dog, I am not prepared to say; the evidence is insufficient for the purpose.

On the question of will-power there is, however, ample evidence for making comparisons.

If the quivering movements of the ears were imperfect voluntary movements, these may be considered about the first manifestation of will in the kitten, and there is nothing to correspond to this in the dog at so early a state.

While attempts to get from the original nest or home took place at an early period in both the dog and the cat, they were more persistent in the latter.

I have given, in considerable detail, the history of the kitten's attempts to get into my book-shelves, etc., and I must repeat that this furnishes to me the most impressive evidence of the existence of a strong will-power, intelligently expressed, that has ever come under my observation in so young an animal of any kind. While I think that the kitten, whose history I have recorded, was above the average in strength of character, if I may so express it, yet, in making all allowance for this, there is still a very large margin in favour of the cat. I doubt if the dog does at any period of his life possess this persistence to the same extent as the cat, and, as in many human beings, this characteristic is associated with unusual physical stamina. The cat's power to live, in spite of its unfavourable surroundings, and her power to resist disease and recover from injury, are undoubtedly greater than in the dog.

The cat is notoriously an independent creature, and in common estimation devoid, or nearly so, of docility; but this very independence and readiness to resent tends, as I have before explained, to cause the cat to be misunderstood. I have, with set purpose, given in great detail my kitten's history, with reference to education in cleanliness, and growing out of this subject alone, a long, and I venture to think, valuable paper might be written on the subject of the education of animals and human beings.

It will be observed that the kitten's instincts were met by placing a sand-pan directly in its path from a box in which it slept to the book-shelves, which it was determined to visit.

From the first moment that its foot was placed in the sand, I noticed that a powerful appeal had been made to the creature's psychic nature,—a new experience engendered a new psychic life,— awakened dormant emotions, tendencies, etc., and these were fundamental. To my mind, this is at the very root of all sound education.

At times, it is true, a little gentle restraint had to be used to prevent the chain of psychic connections forged by these experiences from being broken. But how different the result in this case from that which followed opposition to the kitten's going among the book-shelves. The latter was an instructive thing, the expression of the feline nature to seek retirement in the day-time, and so strong was it, and so supported by will-power and intelligence, that this kitten baffled human efforts in this case to go counter to its nature. I have been accustomed to encourage even young puppies, as soon as they are able to leave their nest, to form habits of cleanliness, but I have no notes on this subject at all so complete as in the case of this kitten, though some will be found in my first paper on the dog.

But I now leave the reader with the diary before him to draw his own conclusions.

The cat can be taught much, but her education must be conducted somewhat differently from the dog's, because her nature is not in all respects like his.

The dog, especially the pure-bred dog, is docility itself. The dog may be forced to obey, the cat cannot. The dog usually delights to obey, or at all events to meet the approval of his master, and he only fails to make this evident when carried away by the force of his instincts. The cat may be coaxed or bribed into docility, but the latter is not a prominent feature in her character.

It is a mistake, however, to suppose that the cat cannot be taught, and taught much, and I think the diary of the kitten, to go no further, shows this clearly.

Certain it is, however, that one will learn more of the cat's intelligence by quiet observation, than by any attempt to form her nature by education, after the manner so successful with the dog.

The tendency of the kitten to arouse in the evening, and display an activity greater than during a large part of the day, is, to my mind, an early exhibition of a fundamental trait in the psychic life of the *Felidæ*.

They are essentially nocturnal animals, and to witness how early this was shown was interesting.

I have noticed nothing like this in puppies, though it must be remembered that the cat is more like her feral congeners, and reverts to a wild state more readily than the dog—in fact, that such reversion is far from uncommon.

In my first paper on the dog, I have called attention to suggestive actions. In the true sense of the word, the cat is perhaps less imitative than the dog, but so great is her tendency to be excited by any kind of motion, that she can, as is well known, be set into activity, with the greatest ease, by a ball, or almost any moving object when a kitten of a certain age.

In this susceptibility the cat is in advance of the dog—in fact, her motor energy is more intense, and her power of correlated movement much greater, but I am inclined to consider that in all this the cat is less imitative than the dog. The behaviour of one kitten has less influence on the others than of one puppy on his fellows.

The individuality of the cat is intense, though it is the individuality of a strong nature manifesting itself by independence rather than great difference in conduct.

As an admirable example of associated reflexes, to which reference has been made in treating of the dog, the history of the sand-pan furnishes excellent examples.

The whole history of the kitten is an illustration that, however strong instincts may be in an intelligent animal, its psychic life is determined by experience, *i.e.* there comes to be almost no pure instincts—instincts unmodified by experience, if such a thing is conceivable, as the language of some writers would seem to imply. Each day of this kitten's life showed me a progress dependent on experience, and the same applies to the dog; but I must add that for the first eight or ten weeks the kitten seemed to get the most out of its experience, though in the case of the mongrel, whose nature, as I have pointed out, seemed after a time to develop with great rapidity, under the impulse of experience, was a rival in this respect with the cat; but that case is exceptional, I must believe.

As regards reasoning, I have in nowise changed the opinions I expressed in my first paper on the dog, and I would apply them with almost, if not quite equal, force to the cat.

Some General Conclusions.

The conclusions that may be drawn from the diaries of the dog and the cat respectively, with certain modifications in some directions, hold for both.

This applies especially to the larger proportion of what is most fundamental, to what is instinctive, and is bound up with the vegetative life of the creature.

Nevertheless, even in some of these fundamentals of psychic life there are differences, *e.g.* in the mode of waiting for and securing prey, differences which appear long before development is complete.

Upon the whole, the cat develops more rapidly than the dog.

The greatest difference between the cat and the dog is in their relations to man and to their own species.

The dog is essentially a social and a gregarious animal; the cat an independent and solitary creature, traits which are early shown.

The dog is docile in the highest degree; the cat to a slight degree, as compared with the intelligence she possesses.

The cat is far in advance of the dog in power to execute highly complex co-ordinated movements.

In both the dog and cat the play instinct is early and highly developed, but in the manifestation of this, the peculiar qualities of each are well exhibited.

In will-power and ability to maintain an independent existence, the cat is superior to the dog.

In the higher grades of intelligence, the wisest dogs are much in advance of the most knowing cats, which is foreshadowed, if not actually exemplified, in the early months of existence.

The nature of the dog, as compared with the cat, tends to beget prejudices in his favour with the mass of persons, in any comparisons as to intelligence, desirable qualities, etc., so that there can be little doubt that in general the dog is over-estimated and the cat under-estimated by the great majority of persons; at the same time, the nature of the dog is much nearer that of man's than is the cat's.

The kitten may amuse, but even a puppy dog touches chords of sympathy in the heart of man that the cat can never reach.

V.—The Rabbit and the Cavy, or Guinea-Pig.

In pursuance of the plan followed in other papers, I shall first give a record of observations on which to base conclusions.

Both common and pure-bred rabbits have been studied, and this rodent will be considered before any comparison is made with the cavy or guinea-pig.

The following notes refer to a litter of common rabbits.

Diary.

1st day.—The animals are found on the first day to be *blind*, the eyelids not yet having opened, and *deaf*.

They are also *naked*, or almost entirely lack hair.

They can move about, but only in a *sprawling*, feeble way.

They lie in a nest lined with the mother's hair, and the slightest movement near causes them great *disturbance*.

A couple are removed from the nest after making observations on them there, and taken into the house for closer study.

The slightest touch, or even a slight puff of air from the mouth causes them much disturbance; they move in an irregular, ill-co-ordinated way, but evidently are greatly affected.

It is especially difficult to get anything near the *mouth* without causing movement, owing, no doubt, in part at least, to the "feelers" in some instances, but not always.

A fly crawling across the head causes jerky movements of the head as a whole, and of the ears.

Irritating liquids, as blistering fluid, iodine, carbolic acid, etc., when held before the nose, cause movements of *face* parts.

Aloes, in solution, and milk placed before the nose give negative results.

When solutions of Epsom salts, common salt, and aloes are placed on the tongue, this organ is protruded in a way suggestive of dislike.

There is in one case even a feeble attempt to *wipe away* the substance with the paws.

A pinch causes violent movements, though there is *no sound* made. In fact, as far as my observations go, young rabbits rarely, if ever, utter any sound.

The *ears* are inclined backward and lie close to the head, as in dogs, cats, etc., at birth.

The animal can manage to *stand* for a moment, after a fashion, but the usual mode of progression is by a sort of *crawling*.

From the first the respiratory movements of the nostrils, so characteristic of rabbits, are shown.

They can also *wipe* the face with the fore-paws—a very characteristic act of the rabbit, or perhaps one should say, of the rodent.

The manner in which they push under one another and *huddle* together, shows how they, like other young animals, are rendered uneasy by cold and quieted by warmth.

They will *not crawl off* a surface, but when they near the edge turn back.

3rd day.—There is a very noticeable increase in size.

Their movements are more pronounced.

The first *scratching* of a surface (in this case, my hand, as one lies upon it) noticed.

This is a highly characteristic act of mature rabbits.

When they are put back into the straw near the nest proper, they push through and get to the other young ones in the nest, evidently attracted by the *warmth.*

Cannot induce them to suck my finger, as puppies will do readily, though a little more inclined to suck the lips or the end of the tongue.

4th day.—*Hair* is now growing over the whole surface of the body.

7th day.—Irritating liquids used before now produce more decided results, and at greater distance.

The evidence that milk is *smelt*, though doubtful in one case, seems clear in another.

Testing as regards taste I have found very difficult in the rabbit, as the tendency to withdraw the mouth suddenly is very strong.

There seems to be no doubt, however, that a solution of aloes produced disgust.

There is a great advance in movements. They are very *quick and irregular*, and may be to one side or the other as likely as forward or backward. Reference is especially made now to movements produced reflexly.

When placed on a table, the young rabbit moves about in a circle, in a half-crawling fashion, and feeling the way, as it were, with the head.

It uses the *paws* fairly well to get rid of anything, *e.g.* a feather put against the mouth.

It is easy enough now to make out the *colour* of the various members of the litter.

The ears are held *back and down* only as yet, still there is an approach to the position of the ears, at times of old rabbits.

When irritating substances are put before the nose the ears are moved, probably reflexly.

They are still both blind and deaf.

9*th day.*—*Twitching*, as in the case of puppies and kittens, is noticed during sleep.

10*th day.*—The eyes of some of the litter are now open.

When I produce a short, sharp sound by a dog-whistle, taking care that they do not feel the blast of air, they move, turn the head to one side, and also move the ears. The effect is less pronounced after two or three trials (exhaustion) though very marked at first.

Can get no clear evidence of vision, nor any eye reflexes, such as winking.

There is such a tendency to *jump about* that one almost leaped from my hands, though the source of the stimulus was not obvious.

11*th day.*—The eyes of some are not yet open.

13*th day.*—A great advance in growth is to be observed within a very few days.

When one touches the nose, or other parts of the head, they almost *jump out* of the nest, reminding me of an animal reacting to a stimulus, when under the influence of strychnine.

Very decided advance made in the *position* of the ears; they are more upright and are moved more readily.

Can get no evidence of vision.

Can get no reflex from the eyes till the cornea is touched, and not very good then.

About the *hearing* there can be no doubt whatever.

14th day.—The *ear carriage* is still better. *Eye reflexes* feebly present.

They are found out of the nest, and go to the mother to suck when distant from her 2 or 3 inches.

15th day.—Since they strike against the side of the nest (board) I am satisfied there is no distinct vision.

They are able to reach the mother when 3 or 4 inches distant.

They seem to *smell* fresh, green food placed in the larger box, with which their nest communicates.

They do not seem to start quite so much when touched as they did.

The rapidity of general *growth*, and especially of the *hair*, is impressive.

There is an obvious advance in smell and taste.

While not certain, I have a suspicion that they begin to distinguish objects. They possibly take notice of shadows.

The first attempt at *eating* made to-day by nibbling at grass put before them.

They now begin to *lope* in the manner peculiar to rabbits.

They also begin to get up on the *hind* legs.

They will by no means creep off any surface on which they are placed.

They now hear a soft whistle at from 4 to 6 feet, as is evidenced by starting movements and ear movements (reflex).

The ears, however, do not move backward or forward but laterally.

These organs are now thinner, held more erect, and better unfolded, so to speak.

16th day.—I think they begin to distinguish objects by *sight*, though it is difficult to demonstrate this.

It is impossible to produce the winking reflex till the eyes are actually touched.

17th day.—It is difficult now to say whether they approach objects through sight or smell.

18th day.—When I sneezed close to their box to-day, they all huddled together, in a *startled* way, in a corner of their nest, in a manner very characteristic of older rabbits when alarmed.

By the manner in which they leave and enter their nests, I conclude that they retain the *memory* of the relations of objects, apart from visual and olfactory sensations.

The mother is now *weaning* her offspring.

One is seen circling around in the box, as if attempting to find the mother by feeling, making it doubtful if distinct vision is even yet possible to them.

To-day they eat clover.

20th day.—They not only eat green food, but *gnaw* at a crust of stale bread.

They jump about in the box, as if *playing*.

They seem to *follow* by smell a green stalk I hold in my hand.

One is so *startled* by my sneezing that it jumped off the top of a barrel on which I had placed it.

The *ears* are now relatively longer and thinner, and are better held.

21st day.—They detect objects (food) by smell at an increased distance.

They are leaping about a good deal, apparently from excess of vitality.

22nd day.—They *follow* my finger as I move it, so that I think there is now no doubt that this is done by sight.

They *eat* stale bread quite well.

24th day.—The young rabbits are removed to a large cage. They now show *fear* when a sudden motion is made, showing that they *see* well—at some little distance at least—and I am under the impression that this was the case two or three days ago.

They now *lope* mostly like older rabbits, *i.e.* they use the two hind legs together—a sort of leaping.

26th day.—They appear now to see objects at a distance of 3 or 4 feet, if not more.

28th day.—They move on the waving of a hat at a distance of 8 feet.

29th day.—Being cool to-day, they are quieter, and huddle together.

30th day.—They seem now to be practically mature, from a psychic standpoint.

Diary of Himalayan Rabbits.

I now give some extracts from notes on a litter of pure-bred rabbits, which may, in some directions, supplement the facts recorded above, and in others furnish grounds for comparisons and contrasts.

2nd day.—A breath of air causes *reflex* movements.

One of the litter is seen to use its hind-leg to *scratch* itself.

3rd day.—They are seen to use the hind-legs to scratch the body, and the fore-legs to *wipe* the face.

4th day.—Already they show *jerkiness* in movement under the slightest stimulation.

6th day.—Any attempt at voluntary movement, if such it may be called, is very jerky.

When put on a barrel top, one of the rabbits turns round, but does not crawl off.

8th day.—Placed one on the barrel top again. When the hand is held near, it will move towards it, attracted by the heat probably—possibly by smell.

10*th day.*—Eyes not yet open.

Green food held within 2 or 3 inches is *sniffed* at.

Hair has now grown on the body to the length of a quarter of an inch.

12*th day.*—*Eyes* of some of the litter now open, but not of all.

The slightest touch with a leaf of green food causes them to *leap* in the nest.

They now move the *ears* and *start* when I produce a moderately loud noise with a dog-whistle. The ears are moved to the extent of one-quarter to one-half inch.

14*th day.*—They now move readily from the nest, which, as in the former case, is situated in one corner, and elevated a few inches, to some other part of the box and back.

Suspect feeble vision, but cannot demonstrate it.

They begin to *nibble* green food.

15*th day.*—Their mode of *progression* is more like that of the mature rabbit, *i.e.* loping rather than walking, in the ordinary sense of the word.

17*th day.*—They are seen not only to *lick* and *scratch*, but to *bite.*

18*th day.*—Begin to poke into *corners* now when placed on the floor.

19*th day.*—There is no doubt that they now *see.*

20*th day.*—One now begins to show the *dark markings* on tail, ears, and nose, characteristic of the breed.

They now *scratch* hard at the wood of their box at times.

21*st day.*—They *eat* oats.

25*th day.*—A fair pad of hair has developed on the foot.

They are now put down with the members of the other litter, twenty-two days older, and act very much as they do.

28th day.—Their hair has grown about as long as that of their parents.

The Cavy, or Guinea-Pig.

The following notes were made on a litter of common cavies, and will serve to mark the contrast between this rodent and the rabbit in the rate of development, etc.

Diary.

The cavies were born between 2 P.M. and 5.30 P.M. in July, and were tested at 9 A.M.—say, after about seventeen hours.

Not only are the eyes *open*, but they *see* well, and when the finger is moved before the eyes the *winking reflex* follows.

Some of them are placed on my study table, and *run* so fast they almost get off the table before being caught.

The *ears* are well *opened up*, and when I whistle moderately they again run almost off the table.

I find it more difficult to demonstrate whether they smell or not than in the case of the rabbit.

When volatile, pungent liquids, like blistering fluids, are placed near, the eyes seem to close, and the same happens with aloes.

Like rabbits, they are born with teeth (in front), and this makes it not very easy to get things into the mouth to test taste.

I am quite unable to decide whether they taste or not. They are tested again at 11 A.M. the same day.

Pungent, volatile liquids, such as used with the rabbit, and aloes do not seem to affect them so much as the rabbits, *e.g.* no sneezing is produced.

When a feather, dipped in a solution of sugar, is placed in the mouth it is *sucked*, but they turn away from similar treatment with aloes.

A couple of them are put into a box 18 by 18 inches, in which there is some salt, some brown sugar, some "peppermint rock" candy, and some camphor.

They *licked* at the salt once but did not repeat this, but went again and again to the sugar or remained by it. They did not remain near the other substances. They were not kept long in this box.

A leaf of lettuce was just after placed before them. They seemed to like to be near it, and very soon began to *nibble* at it.

They *wipe* the face with the paws much like a mature cavy.

Of the three constituting the litter one was from the first much larger.

The larger one was observed to get from its nest to the box, a distance of 2 inches, when not more than three and a half hours old.

2nd day.—At 1.30 P.M. one of the cavies was put into the box occupied by a rabbit. It did not approach or attempt to suck it. I am satisfied it recognised the creature as "strange."

It *eats* sugar from my finger, and follows the finger by sight I think.

The eye must be almost touched before the winking inflex is produced.

3rd day.—*Nibbling* at green food.

4th day.—*Eating* green food.

7th day.—*Eating* food, oats, and stale bread, and they seem, in most respects, to act like old cavies.

8th day.—One is taken to my laboratory for the purpose of brain study. On the way it *squeaks* a good deal.

They eat well, but follow up the mother at times to suck.

10th day.—They now eat as fast and well as mature animals, and in nearly all essential respects resemble them.

There seemed to be little more to record though they were kept to maturity. Other litters were also studied with the same result.

Remarks on the Diary of the Rabbit.

By a comparison of the records for the common and the pure-bred rabbits, it might appear that the latter were, in some respects, in advance of the former, but this is apparent rather than real, I think, as in the common rabbit observations were probably not made quite early enough in some instances.

The marked development of *tactile* sensibility at so early an age in the rabbit is very noteworthy. The creature also responds unusually well, as compared with other animals, as the dog and cat, to *pain-producing* stimuli.

While there is the same uncertainty as to taste and smell at birth, and for a day or two, upon the whole, the evidence is in favour of the rabbit being in *advance* somewhat of the dog and the cat in these respects.

The total absence of voice in the young is in harmony with the sparing production of sounds by the adult.

The movements of the ears, while more marked, are more akin to those of the dog than of the cat.

Movements are developed sooner in the rabbit, and more rapidly become of the kind characteristic of the animal group than in the dog or cat.

The very early date (2nd day) on which scratching was observed, illustrates the rapidity with which co-ordinated movements reach a considerable degree of complexity.

The very early date at which jerky movements are manifest, and which are later represented by that rapid scurrying toward a burrow, etc., is noteworthy. Few animals equal the rabbit in this, and the early development of these movements affords another instance of what seems to be a general law: that those capabilities which are most important in the life of the creature appear early—at all events, as regards qualities essential to the maintenance of existence.

It will be observed that by the 7th day taste and smell are well developed, and the movements of the fore-limbs, as in brushing the mouth, excellently coordinated.

Hearing and vision do not seem to be developed much sooner than in the cat or dog, but hearing especially, as in these creatures, reaches perfection rapidly.

In spontaneous attempts at eating, the rabbit is very much in advance of the dog and cat.

There are very decided physical changes accompanying the psychic ones, many of which have been noted in the diary.

Remarks on the Diary of the Cavy (Guinea-Pig)—The Rabbit and the Cavy Compared.

The cavy shows so clearly, soon after birth, that it sees, hears, smells, tastes, etc., that it would be hazardous to assert that these functions do not exist at birth.

However, I think very close observation convinces one that they all require appropriate stimuli to develop them—that is to say, a cavy does not see, smell, or taste as well during the first hour of its life as it does a few hours later, and marked as is the progress, there is a real development, though the steps towards perfection are rapidly taken.

The contrast with the rabbit—not to mention the dog and the cat—in the condition at birth, and the extreme rapidity with which perfection in all respects is attained, is striking in the highest degree.

During the first five or six days of life the rabbit and the cavy are wide apart, though they both belong to the same great animal group.

After a month the psychic differences are slight, and at maturity they are physically much alike, though the rabbit is probably somewhat higher in the scale.

In the one case the development of body, correlated with a certain psychic status and some peculiarities, takes place *in utero*, in the other case after birth, and that this contrast should be manifest among creatures in many respects so closely allied, both physically and psychically, is especially instructive.

Some excellent observations on the cavy will be found in Prof. Preyer's "The Mind of the Child."

General Conclusions.

The investigations on the rabbit and the cavy illustrate sharp contrasts at birth, and for some time after, in animals that, in mature life, have much in common, both physically and psychically.

The cavy, soon after birth, is able to care for itself, and can maintain an independent existence.

The rabbit at birth is blind, deaf, incapable of any considerable locomotive power, and is, generally speaking, in a perfectly helpless condition.

But this creature attains to a condition of comparative maturity, physical and psychic, within a month, so that it is then quite capable of caring, in all respects, for itself. All its instincts, except the sexual, are in full development about this time or soon after.

In both the rabbit and the cavy, so simple is their psychic life, that there is little to note by way of advance after they are a few weeks old.*

---

* The following account, by Mr T. Mann Jones, shows that under special circumstances the rabbit may show not a little intelligence and "character": "In consequence of the difference I noted in ability and character as between young and old rats, when I was a boy, in 1862, I procured eight young rabbits, so soon as they could really do without the mother. Within a couple of months I saw that the

After the first month of existence comparison with the dog, cat, and allied creatures ceases to be suggestive. The rodents are left quite behind. They seem capable of little education either by man or by nature. In other words, they get little from experience beyond that which strengthens their instincts and emphasises their simple psychic life.

During this rapid psychic development, physical changes of an equally rapid and decided character take place, and are undoubtedly correlated with the psychic changes.

VI.—THE PIGEON—THE DOMESTIC FOWL.

The Pigeon.

So far as I am aware, no investigations on birds of the kind set forth in this paper have been made, except in the case of the domestic fowl and the pheasant. As my observations on the pigeon are the most complete, they will be recorded first.

I have bred a large variety of pure-bred pigeons for many years, and have kept notes on a considerable number of subjects relating thereto, but the following are the most complete consecutive records bearing on the psychic development of pigeons that I have made, and are accompanied with a fairly complete account of contemporaneous physical changes; and I trust that so long an acquaintance with pigeons may be some guarantee of correct observation and interpretation.

---

majority showed distinct 'character.' I selected the two in which this was most marked. One associated himself with my cats and fowl, and protected the young chicks by driving away strange cats. The other associated with the family and children, and its actions to me were those of a very *attached dog*. It appeared to take an interest in everything I did, sat beside me when I was analysing, and moved among my apparatus with the caution so often noticeable in the cat."

My pigeons have been kept in a large, airy, clean loft, have been well fed, and provided with plenty of good water—a most important matter in the case of pigeons.

They have been at liberty to fly out of the loft freely almost every day. As a matter of fact, the state of health in the entire loft has been good.

All of these things are, of course, of importance in interpreting the records that follow.

It seemed to me highly desirable that not only different breeds, but different individuals should be the subject of observation.

I would remind those not familiar with the habits of pigeons that the male and female, unlike domestic animals, pair up mostly for life if not interfered with, so that, speaking generally, a number of pairs may be kept in the same compartment of a loft without crossing, whether they be of the same or of different varieties, when once they are thoroughly mated.

The male and female sit on the eggs in turn, and both take upon them the work of feeding, which consists of disgorging into the mouth of the young water they have swallowed and partially digested or softened food from their own crops.

During the first few days after the hatching of the young, the parents supply a secretion from their own crops, known popularly as "pigeon's milk," and which chemical analysis has shown lately to be a term not wholly inappropriate.

Diary of the Pigeon.

OWL PIGEON.—Hatched out by its own parents, that also fed and reared it.

When born, the *eyes* seem to be closed.

2nd day.—On *touching* the back of the young one, it moves its head and opens its eyes.

3rd day.—It does *not wink* when its eyes are touched, and it is doubtful if it sees.

On touching the bird it moves more than it did yesterday, and now and then it opens its mouth a little.

On removing this $1\frac{3}{4}$ inch from its mate nestling, it *shifts back* again, guided chiefly by the warmth from its fellow, I take it.

It is placed on a perch about $1\frac{1}{2}$ inch wide. It does *not fall off*, but clings to it. When my hand is held within 1 inch, and below it, the bird puts down its beak, touches my hand, and scrambles down into it.

Every time I use the dog-whistle the bird moves its head, which is a proof of *hearing*, for care is taken to exclude the contact of the blast of air from the whistle.

4th day.—It will, when placed on the perch, as before, put its head down a great distance, but does not move from its secure position; but it does creep off into my hand under the same circumstances as yesterday.

5th day.—It *spreads* out its wings when disturbed in any way, or when in danger of falling off the roost.

There is no evidence that the bird can distinguish objects by the eyes.

The winking reflex seems to be wholly wanting.

A solution of quinine and one of sugar are used to test taste, but with no clear results. When blown on, etc., it utters the *sounds* peculiar to young pigeons.

6th day.—The bird can see at a distance of 1 foot to-day, as evidenced by its movements when the hand is passed before it.

Holds its *eyes open* a good part of the time now.

Under the application to the tongue of solutions of sugar and of salt, there is some shaking of the head, which is possibly evidence of *taste*.

7th day.—Clear evidence of *vision* at 2 feet.

If the bird is placed anywhere out of its nest, it moves about in a restless way.

8th day.—Evidence of vision at 4 feet.

9th day.—Can now see objects distant 6 feet.

When the sides of the beak are touched, movements follow, as if preparing to receive food.

10th day.—Objects now seem to be *visible* to the bird at any part of its own compartment of the loft, *i.e.* at 9 to 10 feet.

To-day, for the first time, is witnessed undoubted *defensive snapping*, or pecking with the beak, when the hand is brought near it, whether in its nest or on the floor of the loft.

Feathers are now grown out far enough to enable one to judge of their colour.

11th day.—Growth of feathers and general growth now considerable.

The bird can now walk fairly well.

12th day.—An improvement in walking and feathering noticeable since yesterday.

13th day.—When placed on the floor of the loft, within a few inches of the nest, it attempts to *return* to the nest through some slats that fenced it in. The nest is on the loft floor, and is covered with sawdust only now, though at first there was also a little straw.

15th day.—The bird can now get back to its nest when 2 feet away, and also succeeds in pushing between the slats.

16th day.—The parents have somewhat neglected the feeding of their offspring to-day, as they are preparing to sit again.

17th day.—Better fed to-day. When either parent enters the little compartment in which the nest is, the young one scrambles over a brick that surrounds the nest to *solicit* food.

19th day.—The bird is now well-feathered.

When placed on the perch, it *clings* more firmly than before, and in a different fashion—in fact, more like a mature pigeon.

27th day. — The bird is practically completely feathered.

34th day.—This owl-pigeon is leading a comparatively *independent* existence, and is out of its nest most of the time, and *flying* about the loft, though occasionally fed by the parents, which are sitting on another nest of eggs (two).

Dragoon-Pigeon hatched out and fed by its Parents.

This one will be named B, to distinguish it from another, C.

1st day.—Eyes unopened. The loudest whistle produces no effect. The bird is evidently *deaf.*

Blowing on it causes the bird to *move.* Putting its beak to one's mouth, it opens the former.

When placed on an elevated surface it *does not creep or fall* off, though it moves its beak about, up, down, etc.

2nd day.—Responds to the dog-whistle to-day.

Eyes open, but usually kept closed.

3rd day.—Eyes more completely open, but lids are usually held closed.

5th day.—Keeps its *eyes open* most of the time.

It can now hold up its head pretty well, which is an impossibility in all pigeons just after birth.

6th day.—The eyes are kept open nearly all the time, and the *head* held better—in fact, well.

7th day.—Progressing, but as yet does not peck with the beak at the approach of an intruder.

9th day.—Does snap with the beak to-day, but does not peck.

10th day.—Pecks to-day.

11th day.—Feathers shooting from the skin, but it is difficult to determine the exact colour yet.

12th day.—A great advance in growth and feathering in twenty-four hours.

When one approaches, it *pecks* hard, and uses its wings offensively also.

13th day.—*Colour* of the bird, though one not naturally easily made out, is now clear.

14th day.—When taken from the small compartment in which its nest is, it walks back promptly and well.

15th day.—The *wings* are well covered with feathers.

18th day.—The bird does not peck when approached, but shrinks.

When its *wings* are drawn out it pulls them in again (as did the owl-pigeon on the 22nd day).

20th day.—Shrinks to-day, but when it gets back into a corner of its nest it then pecks vigorously.

24th day.—Behaviour similar to that of 20th and following days.

28th day.—Out of the nest, looking after itself to a large extent, and beginning to *fly*.

34th day.—The *iris* is turning red. Hitherto it has been of a dull, ill-defined colour.

Another Dragoon-Pigeon, C, of different Parentage.

Though the diary was kept in the same way as the previous one, only such parts of it will be given here as serve to bring out something new or to mark individual differences.

1st day.—I could get no definite reaction to tests for taste or smell.

Eyes held closed, but can be opened.

4th day.—Holds its head somewhat differently from B.

When *touched* turns reflexly towards the source of the stimulus.

11th day.—B has, in twenty-four hours, been the subject of a phenomenal growth and advance in physical development generally, so that it now seems days in advance of C.

16th day.—Very *pugnacious* when approached.

20th day.—Strikes vigorously with beak and wings. It desists when placed in the hand, but is as bad as ever when put down again.

22nd day.—When striking at me it got sawdust into its mouth. This was promptly *ejected*, however.

28th day.—Out of the nest and *flying* about as the other one.

Short-Faced Tumblers, D and E.

Unless otherwise stated the notes refer to D.

1st day.—The *eyes* are fairly open.

The opening of the ear is very small.

2nd day.—E can *sit up* fairly when it does not attempt to move, but if it does it sprawls about badly. The head seems too heavy, and its neck too long, but it is a vigorous specimen of this somewhat delicate variety.

3rd day.—When the wing is pricked the head is dipped downward and forward somewhat forcibly, but not towards the side stimulated.

6th day.—The head and neck are well supported.

The eyes are still held open only *part* of the time. The parents sit over the birds at eventide, though very hot in the loft.

7th day.—The eyes held open almost all the time. *Does not wink* even when the eyes are touched.

10th day.—Much growth and general advancement evident.

13th day.—The bird *pecks* when I put my hand near

it. The growth of feathers is just now slight, but that of the body rapid.

14th day—Still more inclined to peck, etc. Can see growth in twenty-four hours.

15th day.—Pecks vigorously. I placed it on one hand, and on extending the other towards it the bird no longer pecks, but uses *voice*, beak, etc., as when its parents approach to feed it.

16th day.—After being on my hand it is placed back in the nest, and then behaves as noted above on the 15th day.

17th day.—It is now growing rapidly, and *feathering* fast, so that the colour can be made out. Though quite pugnacious before, when taken in the hand it grows quiet.

19th day.—The bird is now about one-third feathered. When approached it uses beak and wings as before, but a touch seems to quiet it.

20th day.—The most pugnacious it has yet been.

22nd day.—Still more pugnacious.

24th day.—Can notice a distinct advance in growth and feathering since yesterday.

26th day.—Does not peck, etc., to-day.

27th day.—Pecking again.

29th day.—It is to-day standing on a brick that is beside the nest, and still pecks.

34th day.—It has begun its independent existence to-day, and is *flying* in the loft, picking up grain, drinking water, etc.

Its *iris* has begun to take on the final characteristic "pearl" colour.

Remarks on the Diaries of the Pigeons.

I have intentionally made records on different varieties of pigeons, in order that it may appear to what extent they resemble and differ in their psychic and physical development.

The resemblances are more readily apparent from the perusal of records than the differences, as many of the latter are of a kind readily enough recognised by an experienced observer, but not easily represented by verbal descriptions.

Special pains were taken to ascertain whether the sense of support, as I have called the quality, is present in birds as in mammals.

From the records it will be seen that it is well marked.

The young bird, placed on a perch, feels about, as it were, with its beak for some solid object, and not finding it, remains where it is, but if it touches anything resisting, it begins to move towards it.

The lower an animal in the scale, the more difficult it seems to be to establish the presence or absence of taste and smell at an early period, if I may judge from those of our domestic animals examined by me.

I am unable to speak with certainty as to whether pigeons within the first two or three days possess these senses or not, but that later they do, I have abundant evidence.

Tactile sensibility, and the ability to feel pain, are present just after hatching, as my records show that even a breath of air gently blown on the young pigeon causes it to move.

Additional evidence is furnished by the fact that a gentle stroking tends to quiet the bird.

They are highly sensitive to warmth and cold. One can quiet the most disturbed and pugnacious young one by gently holding the warm hand, a warm cloth, etc., over it.

A single cold day is liable to kill young pigeons if their parents do not sit over them constantly, and sometimes even when they do. The essential vital processes of the body seem to be deranged by cold.

The eyes, though shut for the 1st day, or a greater part of it, can be opened forcibly. Nevertheless, the slit between the lids is at first very small. There is no doubt, in my mind, that pigeons are blind at birth.

The records show that they can soon see, and so early as the 6th day can distinguish objects at the distance of 1 foot.

The diaries give facts which will enable one to note the rate at which progress in this direction takes place. By about the 10th day the pigeon's vision for objects anywhere in a good-sized loft is excellent.

The practical absence of the winking reflex in young pigeons is noteworthy. However, it is not easy to cause mature birds to wink. Moving an object before its head, when the bird is held in the hand, causes movement of the head rather than winking.

The pupillary reflex is, however, soon and well established in birds.

There is no doubt that the newly-hatched pigeon is deaf, but I have found that hearing may be demonstrated on the 2nd day in some cases.

It is very rare, indeed, that pigeons manifest any signs of hostility, etc., when caught up in the hands.

Birds sitting on the nest will sometimes, however, peck vigorously, and the early age at which this pecking or snapping of the beak is exhibited seems to me noteworthy. It illustrates how purely instinctive the matter is. It reminds me forcibly of the hissing of the young kitten, the more so, as both are often evidences of surprise rather than real hostility.

I have not noticed that the newly-born pigeon has voice, but after a few days (five in one case) the bird "squeaks" on the approach of the parents, and especially when being fed.

Later, the voice is used under such circumstances very persistently.

At first the young pigeon can scarcely sit up in any fashion, but in a few hours this is possible, the greatest difficulty being the management of the head and neck, which often fall to one side or forward.

The gradual progress in motor power and co-ordination has been fully noted in the diaries.

By-and-by the young pigeon recognises its own nest when near it, and when alarmed will retreat to it. This is a matter of vision largely, though, as noted in the case of the young rabbits, there may be some sort of memory of distance and direction through tactile and muscular sensibility or otherwise. The subject is obscure and worthy of more study.

So close is the relation between psychic and physical development, that, from the appearance of a bird, one who has observed closely could be able to predict its behaviour; and this seems to me to be undoubted evidence of some sort of correlation between the physical and the psychic. Now and then it will happen that from one pigeon having been hatched a few hours in advance of the other, by its being better able to persistently thrust forward its beak for food to the parents, it fairly starves the other one, or, if not completely, to such an extent that the difference in both physical and psychic development is very striking. Again, owing to innate vigour, one of the two birds in the nest may make a sudden advance, as was noted in the dog, in which case the same result as just referred to follows.

There are many signs of development that appear progressively, such as changes in the shape of the skull and beak, the method of holding the head, the relative proportion of parts, etc.; but, upon the whole, the rate of feathering is a fairly good guide to progress, both physical and psychic.

Though differences between the mature forms of varieties of pigeons, so pronounced as regards physical form, less so psychically, but still real and always present, are obvious to even a superficial observer, it is interesting to note that even at an early date such differences do appear. To illustrate: the Dragoon is a bird of a very bold appearance, and as compared with many varieties, is somewhat wild. It has been spoken of as the "game bird" of the pigeon family. Such characteristics are manifest in the young before they are twenty days old. They peck sooner and more vigorously in the nest. They are shyer of approach, etc.

This cannot be explained by a more rapid development, for several other varieties mature sooner than they do.

Changes in the colour of the iris are as significant, it would seem, as in mammals.

Some Conclusions.

Pigeons, when hatched, and for the first day or two, keep their eyes mostly closed, though the eyelids are not actually united at birth.

They are both blind and deaf when they first emerge from the shell, and for a certain period afterwards.

Perfection of hearing and vision are gradually but rapidly acquired.

Taste and smell cannot be demonstrated at birth, but can be shown to exist some days later.

From the first tactile sensibility and capability for painful sensations exist in exquisite development.

Pigeons, even more than mammals, seem to be sensitive to heat and cold.

The sense of support is very fully developed almost or quite from the beginning.

Voice is present in a few days after hatching.

Motor co-ordinations peculiar to birds, and the power of recognising their own nest, parents, etc., are gradually but well developed.

Physical changes peculiar to this group of birds, and with modifications for each variety, occur at fairly definite periods, are closely related in time, and are probably in some way bound up with psychic changes.

The Domestic Fowl.

This bird is so different in many respects from the pigeon, that I have thought it better to keep the notes I have taken apart.

These observations were made on pure-bred Andalusian chicks, though many others have been studied.

Diary.

Chick A, hatched out some time before daylight, is taken from the nest for the first time, and tested at 4.30 P.M.

It *pecks* three times in succession at a very small speck on the table, and *touches* it each time. It strikes a crumb about a $\frac{1}{8}$ inch in diameter two or three times, and then *swallows* it.

Soon after it pecks at a smaller crumb, and takes it up and swallows it on the first attempt.

It also pecks accurately at a dark *spot* on the table.

Chick B, of about the same age (same hatch), *picks* up bread-crumbs and particles of hard-boiled yolk of egg without missing.

It also pecks at its own *foot* and the *nail* of one's hand.

Chick C, hatched only a little while since, seems feeble, creeps rather than stands, and is soon tire out.

All those tested do, without doubt, *hear* as well as *see*.

They easily follow the hand by the eyes at a distance of 6 inches.

Can get the winking reflexes only when the eyes are all but touched.

A dark-coloured box, a piece of cotton, and the hand are brought near the chicks in succession. The hand alone is followed, showing that it is the *warmth* that attracts.

Solutions of salt and sugar, applied to the tongue, produce no decisive signs of the possession of taste.

Two pigeons—the one a White Pouter, the other a Black Owl—were brought near, to test whether the chicks would show any instinctive fear. They manifested none whatever; on the contrary, they would *nestle* under them.

The birds are tested again about twenty-two hours later.

Previous to the first testing, they had not been from under the hen, and since then they have been under her and nowhere else.

The three chicks now peck well at all that is put before them, as oatmeal grains, canary seed, etc. They peck readily, and touch the objects successfully. The hardest objects are not always taken up at once, however.

Some scales of dried lime-wash from the wall are placed before the chicks. In one case a chick pecks at a scale several times, then gets it into the mouth, but only to *eject* it.

In another case it is not distinctly taken into the mouth and expelled, but is simply picked up and at once *dropped*.

Water is presented to the chicks. They peck at some *drops* on the side of the tin containing the water.

They accidentally get the beak into the water, when drinking follows.

They do not spontaneously put in the beak and drink, either before this accidental result or after, and in this matter they all three behave alike.

They are seen to *scratch* the head with a foot.

At this age another lot of six, which are with a different mother, do *drink* on invitation (clucking) of the mother.

They also eat rather better on her invitation than without it.

They run to the mother from a distance of 4 feet.

Five hours later.—The mother hen drinks, whereupon two of the chicks run rapidly from a distance of 6 inches and *drink*.

One of them *wipes* its beak on the ground.

The hen is later in a box, and cannot be seen by the chicks, yet they move towards her, *i.e.* in the *direction* of the sound she makes.

3rd day, 2 P.M.—They have been fed a couple of times before to-day.

They are now given very small pieces of meat, with which they run off, *peck* it against the ground, and make off from each other, as does a mature hen. One even escapes through wire-netting into the next " run." They are now out of doors in suitable " runs."

One is seen to swallow a piece too large with no more difficulty than a mature bird apparently.

One of the chicks begins to eat lettuce, on which the mother hen is feeding.

A pigeon (the same one used before) is thrown into the run where the hen and chicks are. It flies about a little, and then alights. The chicks did not show the least fear, etc., though the hen attacks the pigeon, uttering a faint sound (danger signal) peculiar to fowls when a bird, as a hawk, flies over them.

9th day.—Feathers shooting out well. Differences in colour very marked.

14th day.—Tried the Black Owl pigeon, as before. No special manifestations on the part of the chicks, nor were there any when the other of the two pigeons before tried was suddenly thrown into the run and fluttered about.

Remarks on the Diary of the Chicks.

Previous to writing the notes on the chicks that were the subject of the present paper, I had observed fowls, young and old, from boyhood.

The brilliant and suggestive observations and experiments of Mr Douglas Spalding had fallen under my eye, and the criticism of his work by so good an observer as Prof. Preyer, determined me to make some special independent observations.

I had the impression that Spalding's statements (*Macmillan's Magazine*, February 1873, referred to also in Romanes' " Mental Evolution ") were somewhat overdone.

My own observations confirm that suspicion, and justify Preyer's criticisms (" The Mind of the Child "), so that I am of opinion that Spalding's statements require revision, though reliable in the main.

Different chicks behave in a way sufficiently unlike to warrant differences of opinion in detail, and one should be on his guard against statements of a too sweeping character. My own observations, etc., on the chick, agree pretty well with those of Prof. Lloyd Morgan on young pheasants ("*Nature*," vol. xl., p. 575).*

It will be seen from my records that even in the same clutch of chickens there are marked individual differences. Thus one may strike a crumb accurately every time it pecks, and pick it up on the first attempt; another misses, or shows great difficulty in getting it into the mouth.

But few remarks are called for in the diary, in view of what has already been published on the chick by others.

I call special attention to the failure of the chicks to be frightened at any time within my records (fourteen days) by the pigeons placed amongst them, in a way that one would have supposed might have called forth any instinctive dread of a rather large flying bird.

My own impression is that chicks do not, in all cases, show fear when the shadow of a bird, as a hawk, passes over them. In other words, instinct is not the hard and fast thing it is sometimes supposed to be.

The sense of support, not referred to by other observers, is well marked.

The chick is very sensitive to cold, though I think less so than the pigeon, except in, perhaps, the case of the most delicate varieties, as pure-bred bantams.

Some Conclusions on the Chick. The Chick and the Pigeon, etc., Compared.

The chick, when it emerges from the shell, or very soon afterwards—certainly within a few hours—can see,

---

* *See* also this author's "Habit and Instinct."

hear, taste, pick up and swallow food, drink, run about, etc.

Its progress is so rapid that in a few days it can lead an independent existence, provided it be protected against cold, wet, etc.

The chick stands to the pigeon in physical and psychic development, in somewhat the same relation as the rabbit to the cavy or guinea-pig.

In all these cases, when full maturity is reached, the psychic difference is not great. The rabbit and the cavy are about on the same mental plane, and so are the pigeon and the fowl.

They all illustrate general laws of development, and the study of these creatures, somewhat low in the vertebrate and psychic scale, seems to me to throw much light on the problems of psychology, viewed not as human psychology alone, but in the broadest possible sense.

THE FUNCTIONAL DEVELOPMENT OF THE CEREBRAL CORTEX IN DIFFERENT GROUPS OF ANIMALS.

In connection with my investigations on the psychic development of young animals,* it seemed important, in regard to the question of physical correlation, to ascertain, in so far as that is possible by experimental methods, at what period the cortex of the brain becomes functionally active. Nothing, to my knowledge, has been done of late years on this subject. I determined, therefore, to give it as complete an investigation as possible. Realising that breadth of investigation was important, as well as thoroughness, the experiments have not been confined to one or two groups of animals, but cover several. This, together with the desire to report only what was thoroughly well determined, has extended these investigations over a long period and involved much labour.

Only those individual animals have been used the exact age of which was known, and, as a matter of fact, most of them were born and kept under my own observation, so that their exact age and, in many cases, their breeding, etc., were known. It is scarcely necessary to state that the experiments were rendered painless.

---

* The author has thought it well to introduce here just so much of this paper as will make the following one somewhat clearer. Both appeared in the first instance in the *Transactions of the Royal Society of Canada*, in 1896.

General Conclusions.

In the dog, cat, rabbit (and in so far as the writer's experiments go, in the rat and the mouse) neither the brain cortex nor the underlying white matter is excitable by electrical stimulation at birth or for some days afterwards.

The cortex is usually not excitable till about the period when the eyes open, though there are exceptions to this rule, most frequent in the writer's experience in the cat, in favour of an earlier date.

The white matter of the brain, just beneath the cortex, is generally excitable, either at an earlier date than the cortex, or with a weaker stimulus.

The reaction for the limb movements is obtainable invariably somewhat earlier in the dog and the cat, and generally so in the rabbit, than those for the neck, face, etc.

Localisation for the cortex, and still more for the white matter, is at first ill-defined, but gradually, though rapidly, becomes more definite.

In the cavy (guinea-pig) the cortex and the white matter beneath are electrically excitable either at birth or a few hours afterwards, and perfection of reaction and localisation is reached in a few days.

Before the brain cortex responds to electrical excitation, ablation of the motor area (centres) leads to no appreciable interference with movements.

The younger the animal, the stronger the current required to produce reaction up to the time that localisation is well established, *i.e.* the weakness of the current required to cause a movement is an indication of the degree of development of the centre in question.

Differences for breeds and individuals exist and

constitute, to some extent, exceptions to the above general statements.

In the above, "cortex" refers to the grey matter in or near the motor area, and "white matter" to the brain substance immediately beneath.

THE PSYCHIC DEVELOPMENT OF YOUNG ANIMALS AND ITS PHYSICAL (SOMATIC) CORRELATION, WITH SPECIAL REFERENCE TO THE BRAIN.

It seemed to me important that psychic and somatic development should be traced contemporaneously, so closely are they related, and in the preceding papers an attempt was made to realise, to some extent, this ideal, but as my researches on the brain were not completed till after the publication of these investigations, I thought it better not to attempt to utilise them at the time. The investigation bearing on the functional development of the cerebral cortex, with special regard to the motor centres, extends to all the groups of animals falling under my studies in psychic development, and is presented in the present volume of the *Transactions*, so that it is now possible to deal with the most important part of the somatic correlation, viz. with the brain. Naturally, I shall draw chiefly from the latter paper, and from those on psychic development for the facts, etc., on which reliance will be placed in attempting further progress, in regard to a more complete correlation of the somatic with the psychic.

No attempt will be made in this paper to discuss somatic correlation in general, as that subject has been treated in the previous papers.'

It would be quite correct to speak of the relations as anatomical and physiological correlation, but as movements are so bound up with the psychic developments of animals, I think it will be more instructive to consider the subject from this point of view, and in doing so, the psychic will be first taken into account.

I.—The Dog.

As soon as a puppy is born, it is capable of cries, crawling, and sucking, and if we except those concerned with the vital or vegetative functions, these about cover all its possible movements. Up to the period when the eyes open there are no new movements. Every one of these can be produced experimentally as reflexes, and the question is: Are they naturally of this character? They improve from day to day, but that is a feature of all reflexes, even the best organised (as swallowing), though it has hardly been adequately recognised.

As pointed out in my paper on the functional development of the cerebral cortex, the latter is absolutely inexcitable at birth, and for a good many days after—indeed, not till about the period of the opening of the eyes—and as I find the white matter also inexcitable at birth, there seems to be no other view possible of these movements than that they are reflex, and that when the brain is called into action, parts lower than the cortex, or even the underlying medulla, in the youngest puppies, must function.

Nevertheless, the animal at this period is progressing, for the improvement of these reflexes implies the more perfect organisation of a neuro-muscular mechanism, which is probably availed of later in all voluntary movements.

In adult life our own movements are often carried out with a perfection in proportion to the degree in which they are reflex, or according to the facility with which higher centres use lower ones, and thus economise psychic energy.

But even so early as the 12th to the 15th day new movements are possible. The eyes have opened, the ears also, and both eyes and ears move, rather reflexly at first beyond doubt, but very soon the puppy moves both eyes and ears voluntarily at times, and still later he fixes the eyes, which is clearly a voluntary act.

It is obvious that there is now an approach to walking (instead of crawling). There are tail movements by the 17th day, and the scratching reflex is excitable. The tail movements are, at this period, almost certainly reflex. Voluntary movements of the tail do not seem to be possible till a good deal later, which corresponds with the well-established fact that the cortical centre for tail movements is not developed till comparatively late.

The barking of the 19th day was probably a reflex, much simpler than such as results later. At this stage, puppies often bark in their sleep, not a common occurrence with mature dogs, though it does take place in dreaming. By the 23rd day, the puppies stand with the paws on the edges of the boards constituting the walls of their pen. This act may be reflex at times, possibly, but on other occasions it is clearly voluntary, and, as they try to get out, we are left in no doubt that they are capable of willed movements, so that by this time, and probably before, there are undoubted voluntary movements. Corresponding with this advance, I have found before the 20th day very distinct cortical localisation for the limbs, head, and face.

Later than this, improvement in reflexes is noticeable, but still more the rapid development of older and the introduction of new voluntary movements, involving more and more complex co-ordinations, and from the psychic aspect the manifest possession of the power to use the machinery of the nervous system and muscles in a way that implies the existence of a growing intelligence and will, and the careful observation of a litter of puppies, as shown in my paper on the dog, will impress both the physiologist and the psychologist with the rapidly-increasing complexity of the life of a young dog, a complexity in which reflex and voluntary movements, instincts, intelligence, emotions, and will blend in varying, but ever augmenting, degrees of intricacy, with all of which the rapidly-developing cortex is correlated, and, as I have endeavoured to show in earlier papers, there is a large amount of somatic correlation over and above that of the brain, which is constant as to period of development, but with variations for individuals and breeds.

The rapidity of psychic development of a terrier, as compared with a St Bernard, is very striking, even within the first six weeks of life, but persists to maturity; and this, I have found, is correlated with a decidedly slower functional development of the cerebral cortex in the St Bernard. The difference in the motor co-ordinations in the latter and the terrier is so striking within the first six or eight weeks of life as to be ludicrous.

II.—THE CAT. THE DOG AND THE CAT COMPARED.

NEARLY all that has been said of the reflexes of the dog applies, of course, to the cat. There are, however, as would be expected, some that are peculiar to the cat, as hissing, which manifests itself at a surprisingly early date in the kitten, long before the eyes open.

As pointed out in my paper on the cat, there is a general and more speedy development in this animal, as compared with the dog, and this holds even for reflexes, *i.e.* they reach perfection more rapidly—in fact, speaking generally, the cat develops faster than even the smaller varieties of dogs, as terriers.

By the 16th day the kitten, specially observed by me, licked its paw. This, under the circumstances, can scarcely be regarded as a pure reflex; certainly, dogs do nothing comparable to this at so early a date. It also scratched its head with the hind leg on the 16th day. Whether this be regarded as voluntary or reflex, it indicates that the cat is in advance of the dog.

Nothing could better demonstrate the more rapid psychic development of the cat than the earlier date at which it steadily follows a moving object with the eyes, or fixes them for some time on a stationary one. In fact, the kitten does this at a time when it is still doubtful if the puppy sees objects, as such, distinctly.

On the 18th day the kitten climbed up the side of its box and tried to get out. Nothing comparable to this occurs in the puppy till a good deal later. It may be said that the history of the cat during the first six weeks of its life contrasts strongly with that of the dog, as regards the more rapid development of reflex movements, the earlier appearance of voluntary movements, and the speedier perfection attained by each, together with the more ready and complete utilisation of experience, the early intelligence, the strength of the will, and the power of attention.

All this is correlated with that earlier development of the cerebral cortex which I have shown occurs in the cat, and there is probably a greater difference than can be made manifest by our crude methods of experiment.

A very marked feature in the psychic development of the cat is the early appearance of the play instinct,* and the perfection of the fore-limb in carrying out the movements necessary for its manifestations. The cat has incomparably better use of the fore-limb at an early date. I have recorded observations on play (with use of the paws) as early as the 22nd day, and, as is well known, the kitten and the older cat have a variety and perfection of movement of the fore-limb never acquired by the dog. This is distinctly correlated with brain development, for, as I have pointed out, movements of the fore-limb are in the cat the first that can be induced by electrical excitation of the cortex, and to this observation my experience leads me to believe there are practically no exceptions, while the case is very different for the dog. Some investigators have expressed the opinion that the fore-limb is also the first to respond in the dog, but this does not accord entirely with my experience. It has occasionally been so in the puppies on which I experimented, but in the large majority the hind-leg responded first. Mongrels and pure-bred animals of different varieties were used. I do not therefore believe that the statement that the fore-leg in the dog is always the first to respond to electrical excitation, can any longer be maintained as a sound generalisation, but it may be, as I have suggested in my paper on the brain, that the truth is, that sometimes the one and sometimes the other limb is the first to react, and that large allowance

---

* The whole subject of play in animals is exhaustively treated by Dr Karl Groos in his "Die Spiele der Thiere," Gustav Fischer, Jena, 1896, which has been translated into English, with additional notes by the German author.

must be made in any general statement for individual and breed differences.

III.—The Rabbit.

Such a creature as the rabbit contrasts in the most marked manner with the dog and the cat.

A rabbit to the last is much more a creature of instincts and reflexes pure and simple, with relatively but little intelligence, all of which is in harmony with its simple modes of existence. Its food is in the wild state usually abundant, and as its escape from enemies is accomplished by swiftness in flight, or by taking refuge in its burrow, there is little in its environment to develop intelligence. With the Carnivora it is quite otherwise. They obtain their food by cunning, stealth, stratagem—it may be concerted action, as in the case of wolves, hyenas, wild dogs, etc.

The ease with which reflex actions are excited on the very first day of existence in the rabbit is striking, and remains a very distinct peculiarity; and on the same day the paws were used to wipe away an offending substance placed in the mouth. On the 3rd day scratching of a surface was observed, an act which has no small part in the burrowing life of rabbits. By the 15th day they eat, and from this date onwards they progress rapidly to perfection of reflex and voluntary action. The early and rapid development of chewing or eating movements, soon associated with the use of the paws to hold food, contrasts in the most decided way with the slowness of the development of good eating movements in the dog, and still more so in the cat. The rabbit's cortex is inexcitable till about the period of the opening of the eyes, on the 9th to the 12th day, and the movements in which the head and face parts are concerned can be induced by electrical stimulation about this time. It is to be specially noted that these movements can be produced in the rabbit, experimentally, almost as soon as those of the fore-limb—in fact, I question whether, in some cases at least, they are not excitable earlier and with greater facility, *i.e.* with a weaker current. With the hind-legs the case is altogether different—in fact, my investigations would lead me to infer that the hind-legs are never related to the cortex in the same way as the fore-limbs. In no case have I been able to establish to my satisfaction the existence of a cortical centre for the hind-legs of the same nature (to put the matter cautiously) as those for the fore-limbs, head, face, etc. The relation between the early, and all but simultaneous, development of the cortical centre for the fore-limbs and head (and face) parts, and the physiological and psychic manifestations of the young rabbit afford one of the most beautiful and striking illustrations of correlation known to me.

IV.—The Cavy, or Guinea-Pig.

The cavy comes into the world able to take care of itself. It can, in a few hours, if not at once, run about quite well, eat, etc. It is at the set-out as far on in the path of development as a rabbit some days after its eyes are open, though in the end there is little difference between these two rodents physically or otherwise.

Corresponding with this advanced physiological and psychic development, the cortex is, as I have shown, excitable at or soon after birth, so that here again there is rendered evident by experiment a close correlation of the kind considered throughout this paper.

V.—BIRDS.

ONE learns how large a proportion of the possibilities, physiological if not psychic, in the pigeon are not dependent on the cerebral cortex, or even the entire cerebrum, by ablation of the latter. Movements, though not spontaneous, are nearly as perfect afterwards as before, and much light is thrown on the nature of reflexes.

I have, after careful investigation, been unable to find any motor cortical centres whatever. The whole cerebral cortex appears to be absolutely inexcitable, except, perhaps, as concerns certain eye movements, and as for these a strong stimulus is required, it is doubtful if they are of cortical origin in the usual sense of the term.

Nevertheless, unless we deny the existence of voluntary movements to the bird—an extreme position—we are landed in physiological difficulties, inasmuch as it has been assumed by nearly all physiologists that the cortex is essential to voluntary movements. The case of the bird seems to me to show that we have much to learn as to the nervous mechanism of voluntary movements, notwithstanding all the investigation that has been given to this subject.

Conclusions.

In the dog and the cat there is a period, extending from birth to about the time of the opening of the eyes, characterised by reflex movements, the sway of instincts, and the absence of intelligence. During this time the cerebral cortex is inexcitable by electrical stimulation, so that the psychic condition during the blind period is correlated with an undeveloped state of the motor centres of the cortex of the cerebrum. The advance in movements, first of the limbs, and later of the head and face parts, together with the psychic progress associated with this, is correlated with the rapid development of the cortical centres for the limbs in the first instance, and later, for the head and face in the period immediately following the blind stage.

This is more rapid and more pronounced in the cat than in the dog, and is correlated with the greater control in the cat over the fore-limbs and with certain physiological and psychic developments characteristic of the cat.

Similar conclusions apply to the rabbit, except that the difference in the rapidity of development of head and face movements is correlated with an earlier organisation of the corresponding cortical centres, and that there is a greater difference between the fore-limb and the hind-limb, with all of which there are special psychic correlations bound up with certain peculiarities of the rabbit's modes of life.

The vast difference in physiological and psychic development of the cavy at birth is correlated with the presence of cortical cerebral centres, readily excited by artificial stimuli, centres which in a few days reach a practically perfect state of development.

The psychic manifestations of the pigeon and the fowl have not the same sort of cerebral cortical correlates as the animals referred to above.

PART IV.

DISCUSSIONS ON INSTINCT.

Prof. C. Lloyd Morgan on Instinct.

To the Editor of *Science*—In an account of a discussion on "Instinct," given in *Science* of 14th February, Prof. Morgan is reported thus: "He described his own interesting experiments with chicks and ducklings, and held that these and other evidence tend to show that instincts are not perfected under the guidance of intelligence, and then inherited. A chick will peck instinctively at food, but must be *taught to drink* [Italics mine]. Chicks have learned to drink for countless generations, but the acquired action has not become instinctive."

In one of a series of papers now in the Press on "The Psychic Development of Young Animals and its Physical Correlation," I have given in detail an account of a study of the pigeon and the chick. It so happens that this very question of drinking by chicks has been especially noted, and I find a record of one observation to the effect that a newly-hatched chick, pecking at the drops on rim of a vessel containing water, accidentally got its beak into the liquid, whereupon it at once raised its head and drank perfectly well in the usual fashion for fowls. Was this by teaching or by instinct?

Later, the chicks seem to peck and drink, sometimes on seeing the mother do so. The act seems to be, in such a case, a sort of imitation, so far as its inception is concerned. But will any one contend that that first act of drinking, referred to above, was other than instinctive? Again, when a chick first drinks, on its beak being put into water, can the act be considered as the result of teaching? Is the chick so intelligent as to carry out an act so complex in such a perfect way, as it does on the very first occasion, as the result of "teaching"? Surely no one will deny that sucking is an instinctive act, yet a newly-born mammal sucks only when its lips come *in contact* with the teat. Is not the case very similar with the chick? The only difference is, that the chick is slower to *recognise* water than food, but as soon as the beak touches water it drinks, and there is no teaching about it. Considering how seldom a fowl drinks, yet pecks all day long at particles of food, it is not surprising that the chick is slower to recognise water (drink) than food. But it is one thing to say that a chick learns to recognise drink, and another to affirm that it learns to drink. The process of drinking is quite as perfect as that of eating from the very first, if not more so, for a chick at first often misses what it pecks at, and fails to convey the object into its mouth in other cases, though it may touch it.

The view that instincts are perfect from the first, and undergo no development from experience, I believe, after much observation, to be as erroneous as it is ancient.

Instinct is never, perhaps, perfect at first, and, so far as I can see, could not be owing to general imperfect development in the animal of motor power, the senses, etc. A young puppy will suck anything almost that can pass between his lips, as a chick will peck at any light spot or object if small, be it food or not. My own records abound in observations that amply prove the position taken, and while my experiments and observations on birds are in the main in accord with those of Prof. Morgan, so far as I know them, I cannot but believe, if I have correctly understood his views, as reported at the New York Meeting, that he has misconceived or overstated the case under consideration.

The subject of heredity is too large to enter upon now. I may say, however, that my researches in Comparative Psychology, and especially in that part bearing perhaps most closely on the question—psychogenesis, do not incline me to believe any the more in that biological *ignis fatuus*—Weismannism.

WESLEY MILLS.
M'GILL UNIVERSITY, MONTREAL.

PROF. MORGAN'S observations agree with those of Prof. Mills and others. A chick swallows water instinctively, but must be taught to drink by example or by accident. The chick might die of thirst in the presence of water, as the sight of the water does not call up the movements of pecking at it, as do food and other small objects. The mother hen replaces natural selection, and the action, though continually practised by the individual, has not become instinctive, because it has not a selective value. Prof. Morgan's argument seems to be satisfactory. If actions which occur but once in the lifetime of the individual (*e.g.* the nuptial flight of the queen bee) are thoroughly instinctive, and others which are practised continually by the individual, do not become instinctive in the race, we can scarcely regard instincts as hereditary habits, but must rather attribute them to variations, fortuitous or due to unknown causes, and preserved by natural selection.

THE WRITER OF THE NOTE.

INSTINCT.

To THE EDITOR OF *Science.*—Some remarks appended to my letter, published in *Science* No. XII., on the subject of Prof. Morgan's views on "Instinct" by "The Writer of the Note," in view of the importance of the subject, are worthy of further consideration.

Before drawing conclusions from observations on domestic animals, it is well to consider similar facts in connection with their wild congeners, especially if such conclusions are of a far-reaching character, and it cannot be too well borne in mind that our experiments are very clumsy imitations of nature in a large proportion of cases.

If food be set down in considerable quantity before newly-hatched chicks, and in a vessel similar to that in which water is usually held, they will be relatively slow to recognise and eat such food, but in a wild state the congeners of the domestic fowl, as grouse, pheasants, etc. do not find food or water before them in such way. Their food is distributed, however, much more like the particles we scatter before the chick than does their water supply resemble that of our methods.

A young grouse would naturally get its water from the dew on herbage, possibly from rain-water that had gathered in little hollows of the ground, surface, etc. And when the birds approach a stream, the surface near is moist or wet, the particles it would naturally peck at would be found up to and beyond the very margin of the water, so that the contact of the beak with water in all these cases would be inevitable, and drinking would come about as naturally as eating.

When "The Writer of the Note" says: "A chick swallows water instinctively, but must be taught to drink by example or accident," the latter term evidently having reference to the observation specially described in my letter, he plainly either misses the real point of my observation or neatly evades it. One might as well say a puppy learns to smell by accident, for, in the case in question, the chick did not swallow water merely, but raised its head like an old fowl and *drank* perfectly well on the very first occasion that its beak had ever been immersed in water (as a puppy sucks when its lips first come in contact with a teat, etc.); and this I take it is what happens in nature. The young grouse in the forest, or even the chick on a grass plot or in a garden, would come in contact with water without any assistance from the mother bird.

The assumption that "the chick might die of thirst in the presence of water, as the sight of water does not call up the movements of pecking at it, as do food and other small objects," is purely gratuitous. It is not primarily so much the sight, but rather the touch of water—inevitable, as I have tried to show, in a wild state—that in the very first instance leads to drinking, though the bird would also peck at shining dewdrops, as my chick did at the drops on the rim of a vessel containing water. With a fair chance, and plenty of water about, in a condition at all resembling that in nature, there is no such thing for a vigorous, hardy chick as death from thirst.

That habits may be hereditary in dogs I have many times observed in my own kennel, during the last eight years, and, without expressing any opinion as to the origin of instincts now, I can see no impossibility in their dating back to habits.

A doctrine which asserts that eating is instinctive, but that drinking is not, is, to my mind, one to marvel at, and is a poor foundation for theories of evolution or heredity.

Comparative Psychology will, I fear, continue to suffer till those who assume to deal with it authoritatively spend more time among animals and less in their studies. A few observations or experiments do not give them insight into the psychic nature of animals, and it were well, I venture to think, if the qualifications of the comparative psychologist, as set forth by Dr Groos, in the preface to his admirable work, "Die Spiele der Thiere," were thoroughly known and believed in by all psychologists. WESLEY MILLS.

M'GILL UNIVERSITY, MONTREAL.

NEWLY-HATCHED CHICKENS INSTINCTIVELY DRINK.

To THE EDITOR OF *Science*.—In the issue of 6th March 1896, appears an excellent and accurate note by Wesley Mills, calling attention to an error of statement made by Prof. Morgan in *Science* (issue of 14th February 1896).

With due deference to "The Writer of the Note," who follows Mr Mills, and who says that Morgan's argument is satisfactory—that " a chick might die of thirst in the presence of water "—I desire to say that this is not my understanding of the case. I have been, during the last thirty-five years, a breeder of fowls as an amateur, and I have given the hatching and rearing of chickens close and continued attention. I have repeatedly placed a shallow water-dish before the bars of the coop in which a newly-hatched brood had been placed the day previous, taken there directly from the hatching nest, and in which they never had food or water offered. Repeatedly, before these small chickens, not twenty-four hours from the shell, and before they had been offered food, I have filled their shallow water-tray, and observed them toddle out to it, peck at it, or at once thrust their bills into it, *to drink at once by uplifting their heads*, as all adult fowls do, the hen never putting her head out from the bars, or showing these young chicks how to do what they instinctively did. I have made the same experiments repeatedly with food, with the same result, *i.e.* that chicks instinctively drink and eat without any example being set by the mother hen.   HENRY W. ELLIOTT.

LAKEWOOD, OHIO, 11*th March* 1896.

THE INSTINCT OF PECKING.

IN discussing Prof. Morgan's lecture on "Instinct," it has several times been stated that chickens pecked instinctively, but had to be taught to drink. There was a note in *Nature* last year, concerning some species of Asiatic pheasants—it may possibly have been the Jungle Fowl—to the effect that the young did not peck instinctively, and did not offer to take food spread before them. The natives seem aware of this peculiarity, and in the particular instance recorded, a native induced the young birds to peck by tapping on the ground with a pencil near the food. They seemed attracted by the sound and movement, and were thus induced to peck at the food.   F. A. LUCAS.

"SCIENCE," 13*th March* 1896.

INSTINCT.

To THE EDITOR OF *Science*.—Having read with considerable interest the discussions under "Instinct," and having noticed the different opinions expressed concerning the eating and drinking of the chick, I thought that perhaps my personal experiments in regard to the matter might be of interest.

About eight years ago I was desirous of studying the chick before and after hatching, and for this purpose I placed about three hundred eggs in an incubator. I shall confine myself to those that were allowed to hatch.

Those that hatched were divided into two groups,— an unhealthy and a healthy group. Those in the first group were fed and given water until they became strong enough to care for themselves. Those in the second group had food and water placed so that they could get them, but they were not fed nor given water, nor were they taught how to secure food and water. No tapping on the dish or on the floor, and no putting of the bill in the food or water was practised. They were left entirely to themselves.

By watching these chicks, I noticed that they would occasionally run over their food and water, and frequently they stumbled in them. If the beak became wet, up would go the head, and the water was swallowed. If food adhered to the beak, some would get on the tongue, and it would be swallowed. In time they seemed to recognise that the food and water were palatable by repeatedly stumbling in them and getting them on the beak, and finally, they *learned* how to secure them, *i.e.* how to pick them up. I noticed that at first they did not know how to pick up, but, after repeatedly trying, they learned how. The majority of these chicks lived and developed.

Now, if we consider the attempt to pick up, from observation I conclude that it was by *instinct*; but if we consider the picking-up, I conclude that it was an *acquired* characteristic.

In conclusion, I might say that at the end of the 3rd day all of the chicks—about fifty—instinctively attempted to pick up, and that at the end of the 5th day they were able to pick up and place the food or water so that it could be swallowed.

J. C. HARTZELL, JR.

ORANGEBURG, S.C., *25th March* 1896.

To THE EDITOR OF *Science.*—Prof. Lucas seems to me to have advanced this discussion on "Instinct" by his reference to a letter in *Nature*, which appeared in vol. lii. p. 30. According to the writer, it is customary for the Asamese natives to "teach" the young Jungle Fowls to peck.

If this be true, what then becomes of Prof. Morgan's distinction?

As a matter of fact, if one observes a good many chicks, he will find that a large proportion of the birds never peck without suggestion (the term "teach" seems objectionable) from the hen or some substitute. The chief value of such facts grows out of their showing that instincts are never perfect, and never of that type once believed in—the unalterable, inevitable, and unvarying—like the rising and setting of the sun; and for such rigid notions the reports of some scientists are in part responsible. It sometimes happened that experimenters in biology, etc., omit the exceptions and report only "good experiments," so that a false view of the case must necessarily arise. Prof. Baldwin seems to adopt Prof. Morgan's views, for he refers to the observation that the chicks drank "only after they had the taste of water by accident, or by imitating the old fowl." Granted—but they also peck only after *seeing* small objects under certain conditions, and there is no instinct that does not require some stimulus in the environment to bring it into action. The mechanism is ready, but it is useless without this stimulus.

If one knew but of those domestic chicks, or those jungle chicks, that peck only on seeing this act, one might speak of a certain imperfection in the instinct of pecking, as, if you will, in drinking; but what I must again object to, is drawing radically different conclusions as to the nature of eating and drinking by chicks, and even building theories of evolution on them.

As I understand Prof. Cope is to reply to Prof. Baldwin's views on "Consciousness and Evolution," through the medium of the *American Naturalist*, I will only remark regarding his discussion in *Science*, p. 438, on "Heredity and Instinct," that, while I find his views very interesting, as illustrations of natural selection, the Lamarckian principle, the influence of environment, etc., they seem, in the main, to fall within the range of principles already recognised by the Darwinians and Lamarckians, though perhaps not adequately. But I fail to see that a single safe step can be taken in explaining evolution either in biology or psychology, if the effects of the environment and of use be ignored; indeed, Prof. Baldwin's very facts and illustrations are, to my mind, only comprehensible by the introduction of those factors; and why there should be such anxiety on the part of many to get rid of factors so obvious, and to substitute for them the biological fatalism and reasoning in a circle of Weismann, is a puzzle to me.

I trust Prof. Baldwin will not insist on coining many new terms, or favour their adoption, as far as evolution is concerned. "Social heredity" is about equivalent, is it not, to social environment, and the entire environment is one into which, as a rule, the animal is born, so why speak of "social heredity"? Technicalities have their advantages, but they often conduce to mental myopia, and hamper the comprehension and progress of truth by binding it up in packages, so to speak—packages which all cannot readily undo.

WESLEY MILLS.

M'GILL UNIVERSITY.

IN Prof. Mills' communications on "Instinct," he seems to have missed the point in the case of each of those criticised—"The Writer of the Note," Prof. Morgan, and myself. In the case of the fowl's drinking, it is not the mere fact that drinking and eating may differ in the degree to which the performance is congenital; the reports seem to show that this varies in different fowl, but that instincts (in this case drinking) may be only half congenital, and may have to be supplemented by imitation, accident, intelligence, instruction, etc., in order to act, even when the actions are so necessary to life that the creature would certainly die if the function were not performed. That is the interesting point.

Then, in criticising me, Prof. Mills accuses me of ignoring the "effects of environment and of use." On the contrary, these are just the facts which I appeal to. By adaptations to the environment and by use, the creature manages to keep alive; other creatures die off; so certain determinate directions of congenital variation are singled out and inherited. Thus phylogenetic variations become determinate, just through these ontogenetic adaptations. This takes the place of the Lamarckian factor. Lamarckism is an "obvious" resort in all cases, of course, but it seems to me so easy, that in many cases it is shallow in the extreme.

But my view is very far from being Weismannism. I reach determinate variations by means of new functions or adaptions which keep certain animals alive to propagate. It is really a new theory, as Prof. Osborn, who has reached about the same point of view, declares. This is also just the value which Prof. Morgan attaches to his observations.

J. MARK BALDWIN.

PRINCETON, 17*th April* 1896.

To THE EDITOR OF *Science*—It seems to me that it would be well to keep the issue with which this discussion started in view, and then the direction in which the truth lies will be clearer. Nothing could be more explicit than the statement by "The Writer of the Note" in *Science* of 14th February, which was this: "A chick will peck instinctively, but must be taught to drink. Chicks have learned to drink for countless generations, but the acquired action has not become instinctive."

In other words, the view that eating is instinctive, and drinking is not, was that taught by Prof. Morgan and endorsed by "The Writer of the Note" in a subsequent communication. Feeling that an important truth was being imperilled, I advanced facts to show that such a view was untenable. This was followed by the recital of additional facts by others, so that it was plain to myself—more so than ever—that such a theory as that first advanced was not sound. I was aware that all three of the writers supporting this view were in accord, constituting a sort of trinity in unity; there was, nevertheless, a great lack of harmony which seemed to be owing to the somewhat important defect that their views were not endorsed by Nature.

Now, to my surprise, Prof. Baldwin claims that I have missed the real point, which he takes to be that an instinct may be only "half congenital," and cites this drinking of chicks; but, according to the above quotation, drinking is not instinctive at all, so that it looks as if the shoe was on the other foot.

In 1894, in a paper read before the Royal Society Canada, on "The Psychic Development of Young Animals," published in the *Proceedings of the Society* for 1895, and a copy of which was forwarded to Prof. Baldwin, I emphasised the conception that instinctive acts are *never perfect* at first, or, as Prof. Baldwin would prefer to say, are only partially congenital, though whether such an expression as "half congenital" is a valuable addition to the English language I doubt. Now, it would be strange that I should alter my own views without noting the change, and miss the point in a matter which I was, I think, the first to emphasise; in fact, I have, in this very correspondence in *Science*, urged this view—the imperfection of instincts. If Prof. Baldwin, and those he professes to interpret, will grant that eating and drinking in chicks are instinctive; that both alike are imperfect at birth; that, congenitally, the chick is in the same condition to all intents and purposes as regards eating and drinking, he will, I believe, be in accord with the facts, and we shall all agree that the much-overlooked imperfection of instincts is well illustrated by the subjects under discussion, but, I should like to add, universal in its application, though in varying degree, the imperfection being in some cases not very obvious to our inadequate observation.

But in discussing evolution I feel that we are on a different plane. Here the appeal to facts is of a much less decisive character.*

I have been trying, since reading Prof. Baldwin's letter in *Science* of 1st May, in reply to my own, to ascertain his real views in regard to evolution, and have some hesitation in deciding whether I really grasp his meaning or not. However, a few concrete cases may make matters plainer. A and B are, let us suppose, two individuals that survive because they can and do adapt to the environment; X and Y die because they cannot; or, in Prof. Baldwin's terminology, A and B adapt to their "social heredity" constituting "organic selection," which is ontogenetic, or affects the individual. But the survival of individuals specially adapted affects the race or phyllum. But surely an individual adapts to an environment ("social heredity") because of what he is congenitally. In the language of evolutionists, this is survival of the fittest, or natural selection, though Prof. Baldwin seems to think he has introduced a new factor in his "social heredity." The name is new, and to my mind objectionable, as there is no real heredity; the idea is not.

Ordinary people express themselves by saying that we become what we are because of "education," "circumstances," etc. We say: "The man is the product of his age."

People tend to believe too much in the power

---

* Although the bearing on evolution of the observations under dicussion was not the principal theme of these communications, it may be stated that, under "Determinate Evolution," Prof. Baldwin has elaborated his views in their most mature form in the July (1897) number of the *Psych. Rev.*, in which article also, reference is made to the opinions of others holding views similar to his own. Prof. Morgan has discussed the subject fully in his "Habit and Instinct."—W. M.

of education, circumstances, etc., and too little in heredity; hence all sorts of cures for deep-rooted evils are ever welcome. But we find that the changes wrought by "social heredity" are very much on the surface, and, in consequence, there may be but little outcome from these effects—possibly none in some cases—in heredity, as ordinarily understood, which does not, however, contravene the Lamarckian or any other well-recognised principle of heredity or evolution. To return to the concrete: A and B have offspring, differing slightly from themselves; the "social heredity" has had little effect, therefore, on the race in the case of the lower animals, much less than in the case of man, possibly, and if the offspring C and D be placed in widely different environments, the slight extent to which they have varied (congenitally) will be all the more evident.

A Lamarckian explains these variations, such as they may be, by the influence of the use and disuse of parts, and evolutionists of other schools in other ways. Prof. Baldwin misapprehends, I take it, the sense in which I employed the term "use" in the phrase in which he quotes from my last letter. The Lamarckian sense was that intended.

I must repeat that, after reading a good deal of what Prof. Baldwin has written on this aspect of evolution, it still seems to me that while he has, with new terminology, set forth old views in a new dress, that there is really no new principle or factor involved. I do not, of course, consider such writing without special value, though it may sometimes be provokingly difficult to understand from the new technicalities employed, for the relative parts played by heredity and environment in the make-up of each individual is an interesting and practically very important problem.

If I have failed to understand Prof. Baldwin fully, and so to appreciate his views at their full value on the score of originality, I regret it. However, it is likely that others are in the same case, and I venture to suggest that the remedy for our denseness, if such it be, is to be found in a specific and concrete treatment of the subject. WESLEY MILLS.

M'GILL UNIVERSITY, MONTREAL.

THE HABIT OF DRINKING IN YOUNG BIRDS.*

To THE EDITOR OF *Science.*—In response to a request that has just reached me, may I ask for space in your columns to say that the statement I made with regard to the habit of drinking in young birds was to the following effect? The chicks that I have observed pick instinctively at any small objects at suitable distance. If a small drop of water be such an object, they will peck at that. But if a shallow tin of water be placed in their run the stimulus of the sight of still water does not evoke any instinctive drinking response. If there be grains of sands or food or other objects at the bottom of the tin, they will peck at these, and incidentally find the water. Sometimes they will peck at a bubble on the brim. Sometimes,

---

* As Prof. Morgan explained, in a communication to *Science*, he refrained from taking an active part in this discussion because he was engaged at the time on his "Habit and Instinct," in which the subject was to be fully considered. Since then that work has appeared, and in acute, philosophical insight, clearness, and general charm of treatment, is equal to anything that has yet come from this able writer's pen. In this work he has also, in the most generous way, acknowledged the contributions of myself and others to the subject under consideration.—W. M.

when one is thus led to drink, others will follow by imitation. No sooner does the beak touch the water than, in the domestic chick, up goes the head, and the instinctive drinking response is shown. I have seen ducklings waddle through the tin repeatedly and not stop to drink, though I had reasons for believing that they were thirsty, for when I dipped the beak of one of them beneath the water he drank eagerly, and continued to do so for some time. On the other hand, a little moor-hen or water-hen, when I quickly lowered it, at about sixteen hours old, into water, drank so soon as its breast touched the surface. It then swam off with instinctive definiteness of co-ordinated leg-movements.

The statement of fact (so far as my observations go) that I made was this: That the sight of still water evoked no instinctive response; but that the touch of water in the bill at once evoked the characteristic instinctive behaviour.   C. LLOYD MORGAN.

INSTINCT AND EDUCATION IN BIRDS.*

THE discussion, first provoked by the note in *Science* of 14th February, relative to the origin of instinct and the inheritance of acquired habitual actions, and the remark of Prof. Wesley Mills (p. 441) that "before drawing conclusions from observations on domestic animals, it is well to consider similar facts in connection with their wild congeners," have led me to make a few experiments upon a fledgling of our common kingbird

---

* By Prof. H. C. Bumpus, in *Science*, N.S., vol. iv., No. 86, 21st August 1896.

(*Tyrannus tyrannus*), captured 2nd July, as it was taking one of its first lessons in flight.

As is well known, the kingbird is exclusively insectivorous, and generally captures its prey on the wing, though it does not refuse insects that may lurk in the foliage, and it may occasionally descend to the ground in pursuit of grasshoppers, whose movements have betrayed their whereabouts. Being thus, in its activities, so different from the omnivorous chick, and belonging, moreover, to the great group of Gymnopædes, or birds which, naked-born, are fed in the nest, we might expect certain differences from the instincts and habits of the precocious, downy chick. Such differences may throw light upon the questions of Comparative Psychology, though, as the material for purposes of generalisation is augmented, they may prove to be variations of no direct suggestive value.

From 2nd to the 11th July the bird, almost incessantly calling for food, was kept in the house and fed from the hand with shreds of meat, moist bread, and a few insects. Water was taken from the wet finger, not as a drop from the tip, but finger and all were seized, the subsequent motions of deglutition being the same as though any large morsel were being engulfed. To the present day (16th July) the bird has utterly refused to accept the pendent drop; nor could it be induced to peck a drop from a leaf or from the surface of any object whatever.

On 11th July I offered the bird a small porcelain dish (such as is used for extract of beef) filled with water. Though hungry, and presumably thirsty, no effort was made towards taking the water, but the dish was repeatedly seized with the same eager fluttering that characterised the general reception of any proffered article, edible or not. (It was noted that the tongue during this act was in rapid motion.) While making an unusually awkward lunge at the edge of the dish the bill was accidentally thrust deep into the water, and quickly withdrawn with an unmistakable air of surprise, followed by an effort to *eat* the water held between the mandibles. The jaws snapped, the tongue could be seen shooting back and forth, and the head, first held horizontally, was only slowly tipped backward, and then not in the way of the chick, described as instinctively perfect, but after the retching method of mouthing and swallowing any object not readily responsive to the contractions of deglutition, and which must needs have the added assistance of the attraction of gravitation.

Though the porcelain dish was afterwards repeatedly offered from 11th to the 16th July, and invariably evoked notes of approval, the bird, in securing the liquid, always bit the edge, and never once dipped the beak beneath the surface, nor drank in the approved method of the chick. The earlier awkward movements, however, were greatly improved through repetition. The substance of the water seemed never to be visually observed, and the empty dish held in the hand evoked the same clamorous approval as when filled with water, and was later recognised, even when accidentally met, though a saucer which had not contained food or water evoked no sign of interest.

On the morning of 12th July, it was noted that if water was allowed to fall from a height, the bird became greatly agitated, opened its mouth, and vigorously struck at the descending drops, and several were swallowed with evident signs of relish. Up to this time, while in my possession, the animal had taken food only when placed by the fingers in the gaping mouth, and had made no effort to pick, selectively, the food from between the fingers, nor had it even changed its position on the approach of food, but had remained in one place, fluttering and incessantly calling until the food was brought to it. On the morning of the following day, falling drops were again struck at and seized, though the bird did not relish the accompanying wetting. At noon the drops were again seized and swallowed. Signs of disapproval of the wetting were shown on the morning of the 14th, and on the morning of the 15th the bird avoided falling water and was content with biting the edge of the dish.

From the above observations I am inclined to agree with Prof Mills that the nature of eating and of drinking are not radically different, and, as the physical condition of substances may pass imperceptibly from solid to liquid, so the physiological processes are practically the same whether the food is solid, pultaceous, or liquid, though I should not attempt to compare too closely the relative perfection of the two processes. I do not, moreover, feel that the first act of drinking is, in its totality, necessarily instinctive. In other words, "when a chick first drinks on its beak being put into water" the act may be considered as, very largely, a result of self-teaching.

The phenomena of eating and of drinking have not, in the discussion, been definitely defined, and there has been some lack of discrimination in the use of the word "swallow." The beak, moreover, is mentioned by Profs. Mills and Lloyd Morgan as the organ, the stimulation of which produces the act of drinking, though Prof. Baldwin attributes the action to the stimulation to the sense of taste.

It seems to the writer that the entire process of eating and drinking should be divided into three parts, viz. (1) seizure; (2) mouthing or mulling; and (3) deglutition. It is only in the first of these that the term "instinct," in the sense of inherited habit, is necessarily used. Baldwin, Mills, and Lloyd Morgan are practically agreed that the young chick seizes instinctively on being stimulated by some small, striking object at a suitable distance. This object may be nutritious, or it may be a feather, a pencil, or a nail-head, a drop of water, or a drop of ink. The mechanism is ready, and the stimulus, properly applied, produces the instinctive mechanical, or, as Lloyd Morgan would prefer, organic action.

The object now held between the mandibles and mulled is subject to the examination, strikingly evident in the kingbird, of the tongue, an organ at the same time tactile, gustatory, and locomotory. It stands at the portal which leads from instinctive to reflex action, and is at once the inspector, reporter, and director of that which first stimulated the eye, and now, through a motor response, has been placed where it may stimulate other special sense organs—taste, touch, and probably smell. It is here that instinctive action becomes guided by individual control, and intelligence begins to act through experience.

The mouth-parts of the young kingbird are large, and the deliberate movements are easily observed. I feel, therefore, that this second, and essential, portion of the process of eating and drinking in the small-mouthed chick may have been neglected or overlooked. Moreover, the process of the perfecting of the action of eating and drinking, through repetition and the guidance of the intelligence is, in the kingbird, comparatively slow, and inclines one, on the grounds of Comparative Psychology, to the belief that the complex act of the chick may be only *apparently* perfect from the first, the successive processes of co-ordination being in the chick much more quickly perfected.

The process in the kingbird, as above detailed, gives at least an opportunity for the more definite limitations of those actions which Prof. Baldwin has, perhaps unfortunately, called half-congenital.

The action of the callow bird in deglutition is probably performed as a reflex on the stimulation of the presence of food in the pharynx. Small fragments upon the beak, and in the anterior portion of the mouth, are not perceived, and do not quiet the almost irritating clamour of the gaping young. The enormous size of the mouth, the thickened "lips," and the bright-coloured concentric markings of the oral walls, make a target, the sensitive centre of which (the opening of the œsophagus) only a most awkward parent could fail to hit. We might argue that the young nestling has not, at first, a definite sense of taste; and actual experiment on the kingbird shows that most unsavoury morsels, when placed in the mouth, are swallowed, though not without subsequent signs of surprise, if not of disgust. It is not, then, difficult to perceive that the young bird, while still within the nest, acquires, as a result of the selective activity of the parent, a taste for certain food. The discriminative exercise of the sense of taste is thus a result of direct tuition. The young cow-bird, whose foster-parent has been a vireo, will doubtless acquire a relish for food very different from that enjoyed by, perchance, its own brother, but the ward of a graminivorous finch.

It may be objected that the orphan chick, selecting food without the discriminative direction of a parent, is not a parallel case with the young kingbird. The bird in my possession was so tame, that when it reached an age comparable with the newly-hatched chick, I could take it into the fields and observe it as it foraged, chick-fashion, for itself. I think that I saw it capture its first insect—I, at least, observed its ability as an insect catcher develop from almost *nil* to expertness. During these excursions, observations were made and data collected for the determination of the following questions: Is there an inherited discrimination in favour of the capture of certain edible insects in preference to others? If unsavoury insects are unwittingly taken into the mouth, are they swallowed? If ejected from the mouth, are there signs of disgust? When unsavoury examples are met a second time, are they avoided?

To the first question I can reply that, at first, all insects were indiscriminately seized. A vile-smelling hemipteron was as tempting as a luscious grasshopper or cricket. Distinctly unsavoury insects (*Tetraopes, Coccinella*) were not touched a second time, except with the greatest caution; though species which were only moderately distasteful (*Lema*) might be taken and devoured, but *without* relish. In one case a large brown ant—the first found—was seized, mulled, and vigorously ejected. The next day the bird was taken to the same tree, and, on perceiving a second ant of the same species, eyed it closely and deliberately, and then shook its head and vigorously wiped its beak, with unmistakable signs of recollection. I mention this particular case, though it is not the only one, to illustrate how quickly the bird was self-taught, for the ant was only one of a dozen different species of insects which were met, and it was so instantly seized that a prolonged visual image was not gained. I might add that the kingbird subsequently refused even to try the edible qualities of a large black ant of a different species, though the bird watched the insect's movements with much interest. Profiting by mistakes, it soon learned to examine critically all strange food, before the tongue should force the contents of the mouth on towards the pharynx.

Can we not, then, conclude that the forcing of acceptable food and drink into the pharynx is not "instinctive," but is the result of a series of satisfactory discoveries of the young bird, which lead up to the placing of the food where it will bring about the stimulation of the reflex centre of the gullet, and the accomplishment of the final act of swallowing?—a series which is intelligently adopted by the bird, and improved by practice.

It is perhaps well, before closing, to revert to the peculiar habit of the bird in snapping at falling drops. From the first, the attention was markedly attracted by flying insects, and any small objects in motion seemed to have a peculiar charm. From this fact I am inclined to think that the seizing of drops was no more than the striking at moving objects, though it is possible that the adult habitually takes water on the wing by seizing falling drops of dew or rain.

INDEX.

ABBOT, C., on the Chipmunk, 57
Advances, intellectual, how made, 22
American Naturalist, account re feigning, 69
Audubon, on the Red Squirrel, 69

BACHMANN, on the Red Squirrel, 57
Baldwin, J. M., on instinct, 285, 287, 289
—— on social heredity, etc., 290, 291
Beaver, power of adaptation in, 41
Bedlington terrier puppies, diary of, 144
Beethoven in music, 41
Bell, R., on the Chickaree, 75
—— on the Mink, 78
Bishop, J. P., on hibernation, 79
Boy, views as to dog's thoughts, etc., 3
Breed, its influence, 34
Butler, A. W., on hibernation, 79
Bumpus, H. C., on instinct and education in birds, 293, *et seq.*

CAT, neglect of study of, 39
Cataplexy, hypnotism, etc., views on, 66, 67, 70
Caution in interpretation, 19, 27
Cavy (or Guinea-pig), the young, at birth, 244
—— contrast between, dog and cat, 244
—— conclusions, general, in regard to rabbit and, 245, 246
—— development of senses in, 244
—— diary of, 244-45
—— ears, at birth in, 241
—— eating in, 242
—— eyes, condition at birth, 241

Cavy, locomotion in, 241, 242
—— Preyer on, 245
—— progress in, 242, 243
—— rabbit, compared with, 244, 245
—— remarks on diary of, 244, 245
—— smell in, 241, 244
—— teeth in, 241
—— taste in, 241, 242, 244
—— voice in, 242
—— wiping the face by, 242
Cerebral cortex, functional development of, in, 264
—— general conclusions regarding, 264
Chick, the diary of, 258, *et seq.*
—— colour of feathers in, 261
—— drinking by, 260
—— feathering in, 261
—— hearing in, 259, 260
—— individual differences in, 262
—— instinctive fear in, 259, 261, 262
—— method of eating meat, 260
—— mother, influence of, on 260
—— pecking at objects by, 258, 259, 260
—— Preyer on, 261
—— remarks on diary of, 261, *et seq.*
—— scratching by, 260
—— sense of support in, 262
—— some conclusions regarding, 262, 263
—— Spalding, D., on, 261
—— swallowing by, 258, 260
—— taste in, 259
—— vision in, 259
—— warmth, effect of, on, 259, 262
—— wiping, reflex in, 259
—— wiping beak by, 260

302 INDEX

Chickaree, significance of wide ranges of, 73
Child and dog compared morally, 25
Chimpanzee, expression of, etc., 42
Chipmunk, habits, etc., of, 53, 55, 57, 59, 72
Cortex at birth, in the dog, 268, 270
—— in the dog and cat compared, 270, 273
—— in the rabbit, 273, 274
—— in the cavy, 274
—— in birds, 275
—— general conclusions regarding, 275, 276
Craig, ——, on the Chipmunk, 72

DARWIN, C., on feigning, on migration, etc., 20, 42, 47, 67
Dawes, on Cocker Spaniel, 33.
—— on horse training, 43, 44
Dog, special cases, 33, 35, 36, 39
Dog and Cat compared, 221, et seq.
—— cat in advance of, in co-ordination of movements, 225
—— cats and, education, 227, 230
—— comparison of the brain development in, 226
—— conclusions, general, as to differences between, 232, 233
—— experience and instinct, effect of former on cat, 231
—— independence in the cat, 227, 229
—— intelligence and memory, 228
—— reasoning in the cat, 232
—— slow development of social instincts in the cat, 227
—— suggestibility in cat and dog, 231
—— will-power in the cat, 228
Dreaming in the lower animals, 37

EARTHQUAKES and animals, 41
Eating, is it instinctive ? 281, et seq.
Education, influence of, 8
Elliott, H. W., on drinking of newly-hatched chicks, 282, 283
Environment, influence of, 8
Evolution and intelligence, 25
Experiments needed, 16

FEIGNING in squirrels, 61, 62, 63, 72
—— its nature, 64
—— and intelligence in squirrels, 73
Ferron, ——, on the dog, 36
—— on the trotting horse, 44
Foundation of the Society for the Study of Comparative Psychology in Montreal, 32, 46

GENIUS and animal performances, 13, 19
Germ of human faculties in animals, 20
Germs and animals, 41
Girdwood, G.P., account of Chickaree, 73
Groos, C., study of play in animals, 49
—— on the comparative psychologist, 282, and Preface, p. vi.
—— on play in animals, 272

HABITS, hereditary, 281
Hall, Marshall, on hibernation in bats, 80, 86, 107
Harris, ——, on the beaver, 41
Heidenheim, on hypnotism, 65
Hibernation, discussion of nature of, 105, et seq.
—— in cold-blooded animals, 79, 82, 108
—— in human beings, D. W. Ross, A. Robinson, and C. K. Clarke, 87, 89, 90
—— in pigs, Millar, 87
—— in squirrels, 57, 59
—— in warm-blooded animals, 80, 81, 82
Himalayan rabbits, 239, et seq.
Homing instinct, 29
—— pigeons, suggestion as to the nature of their feats, 14
Horse, and bears, 44
—— causes that retard mental development in, 43
—— special case, 44
—— training of the, 43
Hydrophobia scare, the veterinary profession and the lower animals, 87, 89, 90

IMAGINATION in animals, 21, 37
Imperfection of instinct, 289
Individual, history of, important, 67
Inference from men to animals, 20
Inferiority of animals in all respects questioned, 15
Intelligence and hibernation, 57
Instinct, to explain animal intelligence, 19

JONES, T. MANN, remarks on kitten, tom-cat, sheep-dogs, etc., 203, 204, 220, 221

KITTEN, the diary of, 175, etc.
—— advance of, 202
—— affection in, 200
—— anger in, 189

INDEX

Kitten, attachment of, 197, 199
—— behaviour towards bird, 195, 199
—— —— catnip, 194
—— —— dog, 195, 197, 200, 201
—— —— fire, 196
—— —— flower-pot, 193, 197
—— —— lap, 193, 195
—— —— mouse, 193, 194, 200
—— —— parrot, 188
—— biting in, 182
—— bookshelves, special experience with, by, 184-88, 204, 205
—— call of "Puss! puss!" effect of, 181, 183, 185, 189, 196, 197, 201
—— catching of tail by, 189
—— catching flies, etc., by, 188, 193, 194, 201
—— caution in leaping, 191, 198, 199
—— choice by, 198
—— climbing in, 181, 182, 190, 191, 192, 194, 195, 199, 200
—— concealment by, 193, 197
—— conclusions, general, regarding, 205, 206
—— condition of eyes and ears at birth, 176
—— covering by, 193
—— crouching by, 190, 195
—— crying in, 183, 186, 187, 188, 191, 192, 197, 199, 201
—— discussion of diary of, 202, etc.
—— dislike in, 178
—— eating, etc., in, 205
—— expression altered in, 197
—— —— of intelligence in, 188
—— fatigue in, 200
—— fear in, 186, 187, 190, 191, 192, 194, 196, 200, 202
—— feeding of milk, 185, 186, 187
—— —— of solid food, 189, 191, 193, 194
—— fondness shown for fish by, 197
—— growth of, 189, 191, 196, 197
—— hearing in, 177, 179, 181 192, 202
—— hissing in, 177, 178, 180, 181, 182, 184, 187
—— influence of environment on, 176, 189
—— Jones, T. Mann, remarks of, 203, 204, 220, 221
—— leaping by, 188, 192, 193, 194, 201
—— licking in, 181, 182, 190
—— law of association as explanation in, 205
—— movements of ears, 178, 184, 202
—— —— of eyes, 185

Kitten, movements of head, 179
—— —— of hind legs, 179
—— —— of tail, 179, 182, 183, 186
—— —— in crawling, 176
—— —— in sleep, 187
—— opening of eyes in, 177
—— out of doors, 197
—— pain, sense, latent period in, 177
—— persistence in, 196, 197, 199
—— physical changes in, 188, 194, 196, 197
—— play in, 181, 182, 185, 188, 192, 195, 197, 203
—— power of attention in, 198
—— pupil, reflex in, 203
—— punishment of, 193, 196, 198
—— purring in, 195, 196, 200
—— quivering of ears in, 178
—— reaction towards dog by, 176, 177, 180, 203
—— reflexes, pupillary, 179
—— —— winking, 179, 202
—— resemblances to mature cat, 198
—— running off, 189, 193, 194
—— sand-pan, experiences with, 186, 187, 189, 190, 193, 197, 199, 204
—— scratching, 180, 198, 200, 203
—— shadows, observation of, by, 183
—— sleeping in, 181
—— smell in, 176, 177, 179, 193
—— sniffing in, 189
—— sociability in, 196, 199, 200
—— stalking in, 189
—— stretching in, 187
—— tactile sensibility in, 176, 179
—— taste in, 176, 179, 197
—— teeth, period of appearance in, 182
—— temperature, sense in, 176
—— toilet-making by mother, 186
—— —— by self, 186, 188, 195
—— use of claws, 189
—— use of mechanical means by, 191, 199
—— uneasiness in, 176, 177
—— vision in, 178-84, 202, 203
—— voice in, 179
—— walking and wandering of, 180, 183, 188, 192, 194, 196, 198, 199, 201
—— watching movements of, 192, 193
—— waywardness, etc., in, 196, 197, 199

LINDSAY, ——, works on Animal Intelligence, 26
Litter, different members of, 7

Lucas, F. A., on pecking in pheasants, etc., 283, 285

Man, a law unto himself, etc., 14
Man's superiority; its basis, 26, 39
M'Eachran, D., intelligence in horses, 73
—— paper by, on a dog, 33
Mental life in the mass of mankind, 10
Metcalf, ——, on dogs, 36
Migration, the explanation of, 30
Millar, F., on imprisoned swine, 87
—— on a pig, 40
—— on a dog, 35
—— J., on a Scotch Collie, 36
Mind in animals, its nature, 17
Minot, C. S., on Comparative Psychology, Preface, p. vi.
Monkey, study of the, 42
—— and dog compared, 8
Moral nature in animals, 21
Morgan, C. Lloyd, discussion of his views on Instinct, 277, *et seq.*
—— letter on "The Habit of Drinking in Young Birds," 292, 293
—— mental attitude of young birds, 49
—— on young pheasants, 262
Motives complex, 11

Newton and space perception, 14

Owl-pigeon, 259, 261

Packard, on origin of the dog, 37
Parallelism in thinking of child and dog, 13
Pease, ——, on black-and-tan bitch, 39
People, some hard to understand, 4
Pig, neglect of the study of the, 40
—— trained, 40
Pigeons, the young—
—— colour of feathers in, 251
—— —— iris in, 251
—— conclusions, general, regarding, 227, 258
—— condition when hatched out, 250
—— conditions under which kept, 246, 247
—— diary of, 247, *et seq.*
—— Dragoon (pure bred), diary of, 250, *et seq.*
—— ears, condition of, 252
—— effect of stroking, 254
—— —— warmth and cold on, 248, 254
—— Environment, effects of, on, 249
—— eyes, condition of, 250, 252
—— —— when hatched, 247

Pigeons, the young, feathering in, 249, 250, 251, 253
—— flying by, 251-253
—— growth in, 252, 253
—— herring in, 248, 250, 255, 256
—— homing tendency in, 249
—— iris, colour of, 253, 257
—— movements, general, 249
—— —— of beak, 250
—— —— of head, 250
—— —— of wings, 248, 251, 252
—— —— reflex, 252
—— —— in walking, 249
—— —— in flying, 250
—— Owl (pure-bred), pigeon, 259, 261
—— pecking by, 252, 253
—— —— at an intruder by, 250, 251
—— —— snapping, etc., in, 255
—— positions of, 256
—— psychic and physical development in, 256, 257
—— progress by, 255
—— pupillary reflex in, 255
—— pugnacity in, 252, 253
—— reflex movements in, 252
—— remarks on the diary of, 253, *et seq.*
—— resemblances and differences in, 254
—— sense of support in, 248, 250
—— sitting in, 252
—— short-faced tumbler (pure-bred), diary of, 252, *et seq.*
—— snapping with beak of, 249, 250
—— soliciting food, 249
—— standing, 252, 253
—— tactile sensibility in, 247, 249, 252, 254
—— taste and smell in, 254
—— vision in, 253, 255, 256
—— voice in, 253, 255, 256
—— winking by, 252
—— reflex in, 248, 255
Prentiss, D. W., discussion of phenomena observed by Czermak, etc., 65, 70
President of American Psychological Association, views of, on psychology, 48
Preyer, ——, on shamming death, 68
—— on chicks, 261
Problems, 37
Psychic development and physical correlation, 267, *et seq.*
—— of young animals, 113, *et seq.*
—— bearing of other studies on, 113
Psychologists' attitude towards the subject of Animal Intelligence, 9
Psychology, in colleges, etc., 48
—— its tendencies, 48

Public mind and Animal Intelligence, 45
Puppy, mongrel, diary of, 207, *et seq.*
—— advantages and disadvantages over pure-bred, 218
—— adaptability in, 212, 213
—— barking in, 212
—— behaviour towards a bone, 212
—— biting in, 211
—— characteristic features, when appearing, 220
—— compared with pure-bred, 215, *et seq.*
—— comparison of intelligence with that of St Bernard, 219
—— concussion, effects on, 209
—— conditions under which kept, 207
—— crawling in, 209
—— discussion of diary of, 213, *et seq.*
—— eating by, 211
—— effect of mingling with others of its own kind, 216
—— environment, effect of, 209, 212, 213, 217
—— gnawing in, 210
—— growling in, 210
—— hearing in, 208, 210, 212, 216
—— intelligence in, 212
—— investigation by, 211
—— locomotion in, 210, 211, 212
—— movements in, of ears, 208, 217
—— —— eyes, 209
—— —— hind-legs, 209, 212, 217
—— —— jaws, 210
—— —— tail, 209, 212, 217
—— muscular sense in, 214
—— opening of eyes in, 208
—— perseverance of, 212
—— physical changes in, 209, 212, 220
—— play in, 211, 212, 216
—— pure-bred compared with, 215 *et seq.*
—— reflex effect through ear, 217
—— resemblances to mature dog, 212, 213.
—— sense of pain in, 207, 213
—— sense of support in, 207
—— sexual feeling in, 213
—— shaking of head, effect of, in, 211
—— shyness in, 211
—— sociability in, 211, 212
—— smell in, 207-10, 213, 214
—— starting in, 210
—— sudden development in, 219
—— taste in, 207-9, 213
—— temperature sense in, 207, 213
—— teeth in, 209, 211
—— vigour of, 207

Puppy, mongrel, vision in, 209, 210, 211
—— voice in, 207
—— weaning of, 209
—— winking in, 211
Puppies, pure-bred—
—— anger in, 159
—— are teats found by smell? 148
—— association, mental, in, 133, 139
—— attention in, 133, 161
—— barking in, 127, 129, 134, 139, 145
—— biting and chewing by, 145
—— changes, physical in, 131, 142, 145, 168, 169, 170
—— chewing in, 128, 129
—— concussion, effect of, in, 144, 154
—— co-ordination of movement in, 145
—— crying, crawling, sucking in, 113, 132, 147, 148
—— differences in breeds of, 170
—— differentiation, sexual, in, 144, 145
—— dreaming by, 162
—— drinking by, 127, 128
—— environment, influence of, 164
—— exhaustion in, 124
—— experience, effect of, 143, 166
—— expression, change in, 130
—— eyes, when open, 120
—— fatigue in, 161
—— fear in, 129, 130, 134, 136, 144, 145, 158
—— feeling in, 119
—— general conclusions regarding, 172-74
—— growling in, 125, 128, 137, 143, 144
—— habits, barking, crying, etc., 141
—— hearing, advance in, 125, 137
—— —— discussion of, in, 153
—— —— perception of direction of sound, etc., 127, 130-34, 137
—— —— reflex, 121
—— —— special tests of, in, 125
—— humour in, 161
—— hunger, when felt, 126
—— imitation in lapping water, 136
—— —— in general, 138, 139, 163
—— individual differences in, 124, 137
—— individuality in, 131, 166
—— interest in surroundings, 137
—— investigation by, 138, 140
—— irritation shown, 128, 131, 145
—— jaws and front legs, use of, 121
—— lapping in, 127, 128, 133, 136, 145

Puppies, pure-bred; latent period of pain sense, 144
— licking in, 120, 124, 125
— lip-call, effect of, 126, 161
— movements, when sucking, 129
— — of hind-limbs, 129
— — of head, ears, etc., 129, 140
— memory in, 142, 159
— muscular sense in, 149
— new surroundings, influence of, 118, 120, 130, 145
— pain, sense of, in, 148
— periods of development in, 167
— physical correlation, 168, 170
— play in, 122, 123, 126, 128, 131, 136, 145
— — remarks on, 155
— pleasure in new straw, 133
— — on seeing human being, 135
— reasoning in, 165
— recognition in, 160
— reflexes, associated in, 166
— — in hearing, 121
— — in scratching, 124
— — in winking, 121, 124, 134
— retiring-place of, 134, 142
— scratching in, 127, 132, 136, 141, 144, 156
— similarity to the mature dog in, 163
— smell in, 119, 120, 122, 123, 129, 133, 144, 151
— suggestive action in, 163
— tactile sensibility in, 149
— tail movements, 120, 123, 130, 132, 135, 144, 145, 157
— tails, wagging and position of, 117
— — use in play, 128
— taste in, 131, 134
— — and smell, 151, 152
— teeth in, 130
— temperature sense in, 121, 149
— the mysterious, 166
— uneasiness when removed from pen, etc., 118, 120, 130
— vision in, 121, 123, 128, 130, 131, 133, 152
— voice in, 158
— whip, behaviour towards, 141
— where kept, 117

QUALIFICATIONS to understand animals, 15

RABBIT, the young—
— behaviour when placed with others of older litter, 240
— condition of skin, 234, 235, 236, 240, 241

Rabbits, the young, condition of feet, 240
— co-ordinated movements in, 243, 244
— diary of the common, 233, et seq.
— — pure-bred Himalayan, 239, et seq.
— early development of taste and smell in, 244
— ears, position, movements, etc., of, 234, 236, 237
— eating in, 237-40
— excitability in, 239
— effect of warmth on, 235, 239, 240
— eye reflexes in, 237
— fear in, 238, 239
— growth in, 235, 236
— hearing in, 233, 236, 237, 240
— hearing and vision compared with these senses in dog and cat, 244
— investigation by, 240
— jerky movements in, 243
— locomotion in, 234, 237, 238, 239, 240
— loping in, 237, 239
— movements in, 234, 235, 236, 239, 240
— — of ears, compared, 243
— — generally compared, 243
— memory in, 238
— opening of eyes in, 236, 240
— pain, sense in, 243
— physical and psychic changes in, 244
— play in, 238
— psychical maturity in, 239
— remarks based on the diary of rabbits, 243 et seq.
— scratching in, 235, 239, 240
— smell in, 234, 235, 238, 240
— tactile sensibility in rabbits, 243
— taste in, 234, 235
— taste in rabbit as compared with dog and cat, 243
— twitching in, 236
— vision in, 233, 237, 238, 239, 240
— weaning of, 238
— wiping face, 235, 239
— voice, absence of, 243
Reason in animals, 21
Right and wrong, recognition of, by animals, 24
Robinson A., on hibernating man, 88, 89, 90
Rodents and plasticity of habits, 55
— mimical appreciation in, 60
Romanes, G. J., reference to works of, 26
— influence of, 49

Romanes, G. J. on feigning, 67, 68, 69
Ross, D. W., on "Sleepy Joe," 87
Rule in interpretation of Animal Intelligence, 11
Ruttan, R. F., analysis of urine of lethargic woman, 102

SCALE in intelligence, 23
"Sense experience" in animals, 9
Senses in the lower animals, 23
Shakespeare, sources of his power, 5
—— and Scott in literature, 29
—— on effects of fear, 68
Shamming death, 65, 66, 67
Sheep and storms, 32, 40
Sick animals and Animal Intelligence, 47
Simpson, ——, on the dog, 38
Singing rodents, 59, 60
Society for the Study of Animal Intelligence, founding of, 46
Spalding, D., on Chicks, 261
Squirrels, Black and Red, compared, 74
—— entering traps, 53, 55
—— flying, and sneezing, 56
—— —— intelligence in, 56
—— —— nocturnal, 58
Status of Comparative Psychology, 49

St Bernard puppies, diary of, 117, 144
Studies, detailed, 50
Study, of animals, how to, 35
Summary, imperfect, of Author's views on Animal Intelligence, 15

THINKING, different meanings of, 9
Torrance, ——, on sheep, 40
Training in animals, 23, 33, 35
Turkey buzzard, and feigning, 69

UNDERSTANDING of animals, 5
Unsolved problems, 29

VETERINARY profession, and Animal Intelligence, 30
Views, unscientific, as to man's origin and relations, 18

WATER, aptitude in finding, 43
Weismannism, 279, 286
Woodchuck, hibernation in, 81, 82, et seq.
Words, their understanding of, by animals, 33, 34, 38, 39
"Writer of the Note" on Instinct, 279, 282

YATES, W., on hibernation, 80, 81, 82

www.ingramcontent.com/pod-product-compliance
Lightning Source LLC
Chambersburg PA
CBHW022023240426
43667CB00042B/1079